Next Generation Energetics

2nd International Explosives Conference

Next Generation Energetics

2nd International Explosives Conference

Edited by

Jacqueline Akhavan
Cranfield University, UK
Email: j.akhavan@cranfield.ac.uk

ROYAL SOCIETY OF **CHEMISTRY**

The International Explosives Conference (IEC-2024) took place in Edinburgh from 18–20 June 2024. This publication provides a record of the proceedings for the oral presentations and display posters showcased at the event.

Special Publication No. 259

Hardback ISBN: 978-1-83767-173-1
PDF ISBN: 978-1-83767-755-9
EPUB ISBN: 978-1-83767-754-2

A catalogue record for this book is available from the British Library

The Royal Society of Chemistry is a charity, registered in England and Wales, Number 207890, and a company incorporated in England by Royal Charter (Registered No. RC000524), registered office: Burlington House, Piccadilly, London W1J 0BA, UK, Telephone: +44 (0)20 7437 8656.

For further information see our website at www.rsc.org

For general enquiries, please contact books@rsc.org

For EU product safety enquiries, please email books@rsc.org or contact Royal Society of Chemistry Worldwide (Germany) GmbH, Römischer Hof, Unter den Linden 10, 10117 Berlin.

Printed in the United Kingdom by CPI Group (UK) Ltd, Croydon, CR0 4YY, UK

Preface

Jacqueline Akhavan

This is the Proceedings of the second International Explosives Conference which was held in Edinburgh, Scotland. The conference provided an opportunity for scientists and practitioners from the UK to share their research with the rest of the world in the next generation energetics. The success of the conference was due to the abundance of early careers and the networking opportunities.

Sixteen peer reviewed papers are presented in these Proceedings covering (i) synthesis, characterisation and diagnostics, and (ii) artificial intelligence, simulation and modelling.

1 IEC-2024

Following the success of IEC-2022, IEC-2024 took place in the John McIntryre Conference Centre at Edinburgh University as shown Fig. 1. There were 216 delegates from 21 countries. 73% were from the UK, 12% from Europe, and the remaining from the rest of the world.

Fig. 1 Details of the IEC-2024.

Centre for Defence Chemistry, Cranfield University, Defence Academy of the United Kingdom, Shrivenham, SN6 8LA, UK. E-mail: J.Akhavan@cranfield.ac.uk

2 Keynote speakers

Four keynote speakers were invited from the UK, Europe and the USA. The first keynote speaker was Yogi Gupta from Washington State University. His presentation was on Shock Compression and Detonation of High Explosives: Scientific Challenges and New Opportunities as shown in Fig. 2.

Carole Morrison from the University of Edinburgh was the second keynote speaker and gave her presentation on The (Other) Big Bang Theory: Predicting Impact Sensitivities for Energetic Materials. The third keynote presenter was Carole Rossi from LAAS-CNRS in France who spoke on Nanoenergetic: Paving the Way to Smaller and Smarter Pyrotechnics, and the final keynote presenter was Felix Plasser from

Fig. 2 Yogi Gupta presenting his keynote lecture.

Loughborough University who spoke on Machine Learning Detonation Parameters from Quantum Chemistry Data.

3 Prizes

The Armourers & Braziers Gauntlet Trust kindly sponsored six prizes: Three for the best presentations and three for the best posters.

The first prize for the best presentation was awarded to Carole Morrison from the University of Edinburgh for her keynote presentation. The second prize was awarded to Anna Stammers from Roxel for the presentation on Composite Rocket Propellant Manufacture: A Comparison of RAM and Planetary Mixing. The third prize was awarded to Ashfaq Afsar from the University of Edinburgh on Synthesis, Characterisation and

Fig. 3 The poster session.

Energetic Performance of a New Copper Compound Based on 3,4,5-Trinitro-1*H*-Pyrazole.

For the poster session the first prize went to Barış Edis from Roketsan Missiles Inc in Turkey on 'Effects of slit plate design on mechanical properties of castable PBXs'. The second prize was awarded to Natasha Boulton from the University of Edinburgh on 'New routes to energetic polymers', and the third prize was awarded to Harvey Newman from the University of Edinburgh on 'Towards finding lead-free ballistic modifiers in double based propellants using computational modelling'.

A photo of the poster session is presented in Fig. 3.

4 The conference proceedings

Sixteen papers were submitted for peer review by the Editorial Committee which consisted of

- Professor Colin Pulham, University of Edinburgh
- Professor Andy Tyas, University of Sheffield
- Dr Peter Portius, University of Sheffield
- Dr Bill Proud, Imperial College London
- Dr David Williamson, University of Cambridge
- Professor Jacqueline Akhavan, Cranfield University

Details of the peer reviewed papers published in these Proceedings are presented in Table 1, and details of all the oral presentations and posters given at the conference can be found in Tables 2 and 3 respectively.

Table 1 Details of the papers published in these Proceedings (IEC-2024).

Paper No	Author(s)	Title of paper
Section 1 Synthesis Characterisation and Diagnostics		
1	Ashfaq Afsar, Xiaojiao Liu, Patrick McMaster, Carole A. Morrison, and Colin R. Pulham	Synthesis, Characterization, and Energetic Performance of a New Copper Complex Based on 3,4,5-Trinitro-1*H*-Pyrazole
2	Alice Mintoff, Colleen Williams Christopher Stennett and Christine McGee	The Growth and Shock Testing of Large Single Crystal Energetics and Analogues
3	Gareth Collett, Jacqueline Akhavan, Rachael Hazael, and Richard Critchley	Calculating the Heat of Explosion for Non-Ideal Explosives
4	Francesco Saverio Amendola, Silvana Grossi and Simona Pergolini	Degradation Kinetics of Sea-Dumped Decoy Flares
5	Robert Cargill	Computed Tomography of a Polymer Bonded Explosive
6	Joseph Henry Nelhams	The Effects of Surface Finish on the Signal Return for Het-V Experiments Using Copper

Table 1 *(Continued)*

Paper No	Author(s)	Title of paper
Section 2 Artificial Intelligence, Simulation and Modelling		
7	Jan Langer and Moritz Heil	Prediction of Material Properties of Energetic Materials Using Machine Learning Methods
8	Michael L. Hobbs, Michael J. Kaneshige and William W. Erikson	UCM/MMP Cookoff Models for Explosives Containing RDX
9	Adam Schofield, William Belfield and John P Curtis	Monte Carlo Analysis for Hot-Spot Burning in Generalized Polygons
10	Ajen Limbu, Elliot Tam, Mark Whittaker, Simone Falco, Daniel Eakins, A. Klomfass, Andy Tyas, T. Lodge, D. Farrimond and Daniel Pope	Expansion of High-Fidelity Physics Capabilities for Simulating the Detonation and Behaviour of Explosive Materials
11	Keith R. Coffee*, Jason S. Moore, Ana Racoveanu, Keith D. Morrison, Alan K. Burnham, Batikan Koroglu, Gregory L. Klunder[a] and John G. Reynolds	TATB Thermal Decomposition: An Improved Kinetic Model for Explosive Safety Analysis
12	Sandeep S. Tambera, Claire J. Taylor, Declan H. Williams, Javid Hamidc, Alison Beardah, Justin Fellows, Hugh Patrick, Elizabeth Swann, William Payne, and Charles J. M. Footer	Using AI for the Discovery and Synthesis of Advanced Energetic Materials
13	Simone Falco, Liam Smith, David Chapman, Ajen Limbu[b] Mark Whittaker, Daniel Pope, Nik Petrinic, and Daniel Eakins	Numerical Framework for the Optimization of HVRB Parameters from Experimental Measurements
14	John Philip Curtis, Adam Schofield and William Belfield	Analysis of the Burning of Spheres and Cubes
15	Thomas Scott	Analysis Technique for Reconstructing 2D Cylinder Wall Expansion
16	Alan Williams, Judith Brown and Michael Kaneshige	Frictional Heating Measurements in PBX 9501

Table 2 Details of the presenters at IEC-2024.

Name	Title of oral presentation
Session 1 – Simulation and Computing	
Professor John Curtis, AWE, UK	Analysis of the burning of spheres and cubes
Dr Michael Hobbs, Sandia National Labs, USA	UCM/MMP cookoff models for explosives containing RDX
Dr Jonathan Rougier, AWE, UK	Opportunities for analysis and visualization of detonator experiments using modern statistical methods
Dr Andrew Parker, Fluid Gravity, Engineering Ltd, UK	STASIS calculations of expansion states for detonation products and fits to JWL equations of state
Dr Simone Falco, University of Oxford, UK	Numerical framework for the optimisation of HVRB parameters form experimental measurements
Ajen Limbu, Dstl, UK	The expansion of high fidelity physics capabilities for simulating the detonation and behaviour of explosive materials

Table 2 *(Continued)*

Name	Title of oral presentation
Elliot Tam, Dstl, UK	Recent improvements in Dstl's capability for modelling UNDEX and associated structural response
Charlie Ellcome, AWE, UK	Developments on the analytical modelling of the pinch of hollow explosive cylinders
Dr Rachel Schwind, University of Edinburgh, UK	Updating kinetic model correlations for modelling linear burning rates: The influence of surface temperature of the pressure dependent behaviour of nitromethane
Session 2 – Synthesis and Manufacture	
Dr Ashfaq Afsar, University of Edinburgh, UK	Synthesis, characterisation and energetic performance of a new copper compound based on 3,4,5-trinitro-1*H*-pyrazole
Ruaraidh Wells, NiTech Solutions Ltd, UK	Continuous crystallisation of energetic materials using continuous oscillatory baffled crystalliser
Professor Thomas Wirth, Cardiff University, UK	Flow electrochemistry for Amine functionalization
Elizabeth Francois, LANL, USA	Improved Synthesis of diaminoazoxyfurazan (DAAF) with a manufacturing mindset
Anna Stammers, Roxel, UK	Composite rocket propellant manufacture: A comparison RAM *vs.* planetary mixing
Dr Licia Dossi, Cranfield University, UK	Foundation of a new family of azido binders based on polymeric cyclodextrins
Qi Feng Chan	High pressure structural study of ammonium dinitramide
Session 3 – AI and Machine Learning	
Jan Langer, Fraunhofer ICT, Germany	Prediction of material properties of energetic materials using machine learning methods
Jonathan Miller, Cranfield University, UK	An overview of KIVMesh, a new computational pipeline for datamining computed tomography datasets of energetic materials
Dr Patrick Kimber, Loughborough University, UK	Machine learning properties of energetic materials from quantum chemistry calculations of single molecules
Sandeep Tamber, QinetiQ, UK	Using AI for the discovery and synthesis of advanced energetic materials
Keith Coffee, LLNL, USA	TATB thermal decomposition: An improved kinetic model for explosive
Session 4 – Space and Other	
Alan Williams, Sandia National Labs, USA	Frictional heating measurements in PBX 9501
Brigadier Gareth Collett, Cranfield University, UK	Calculating the hear of explosives for non-ideal explosives
Captain Francesco Amendola, Italian Air Force, Italy	Degradation kinetics of IR decoy flares in seawater
Alice Mintoff, Cranfield University, UK	The growth and shock testing of large single crystal energetics and analogues
Akachai Khumsrii, University of Edinburgh, UK	Cocrystallisation studies of ammonium dinitramide

Table 3 Details of the posters at IEC-2024.

Author(s)	Title of poster
Barış Edis, Rokestan, Turkey	Effects of slit plate design on mechanical properties of castable plastic bonded explosives
Adam Schofield, William Belfield and John P Curtis, AWE, UK	Monte Carlo analysis for hot-spot burning in generalized polygons
Mr Bryn Nicol, Cranfield University, UK	Comparative nitration of bacterial and vegetal derived cellulose
Janset Sine Şekercİ Şanli and Alper Sevinç, Rokestan, Turkey	Effect of ap content on the mechanical, rheological properties and performance of thermobaric PBXs
Dr Bonny Victor, EPC UK	Hydrogen peroxide emulsion (HPE) next generation bulk explosives for civil applications potential and challenges
Dr Bradley Westwater, Sheffield University, UK	Synthesis and characterisation of novel transition metal azido(nitro)triazolates
Thomas Scott, AWE, UK	Correcting cylinder test PDV data with angled probes
Joseph Nelhams, AWE, UK	Effect of surface finish on PDV back reflectance
Dr Maxime Puyo, CEA, France	Towards a better understanding of composite energetic materials, a methodological study of HMX & LLM-105-based explosives
Stephen Spice, University of Edinburgh, UK	An investigation of structure-property relationships in polyvinylnitrate
James Sedgwick-Wilde, MOD, UK	Failure of tactical solid rocket motors
George Tucker, University of Wales Trinity Saint David, UK	Scaling controlled fragment casings
Dimitrios Samaras, Oliver Blackman, Matthew Maisey, Soraia Pimenta and Maria Charalambides, AWE, UK	Quantitative analysis of granular explosives through examination of compaction manufacturing process
Alonso Romero Jabalquinto, Nathalie Mai, and Philip P Gill, Cranfield University, UK	Oxygen and water effects on the decomposition of double base propellants
Romuald Van Riet, Royal Military Academy, Belgium	Detonation simulant of TATP as donor charge of tertiary explosives
Jennifer Mungham, Dstl, UK	Pyrometry for energetic events
John Rosener, LLNL, USA	Using differential scanning calorimetry for chemical compatibility analysis of high thermal stability explosives
Dr Jeffrey Barber, Transportation Security Laboratory, UK	Loaded waveguide dielectric measurements of powdered explosives
Keith Morrison, LLNL, USA	New thermal decomposition reaction for TATB
Mr Jonathan Miller, Cranfield University, UK	KIVMesh - A new computational pipeline for studying energetic materials
Natasha Boulton, University of Edinburgh, UK	New routes to energetic polymers
Andrew Claydon, Cambridge University, UK	Studies on resonant acoustic mixing (RAM) characterisation and processing
Oliver Blackman, Dimitrios Samaras, Soraia Pimenta, Paul Ryan and Maria Charalambides, Imperial College, London, UK	Developing a three-dimensional micromechanical model of a polymer bonded composite with representative formulated microstructure
Jacob Brinkman, Pantex, USA	Investigation of machine response of inert PBX 9501 and PBX 9502 surrogate materials
Simon Barreståll, Henrik Skifs and Lukas Esperi, Eurenco, Sweden	Thermostable explosives development
Mr Robert Cargill, MOD, UK	Micro computed tomography of a polymer bonded explosive
Mr Will Greenwood, Sheffield University, UK	An organic clathrate with energetic potential

Table 3 *(Continued)*

Author(s)	Title of poster
Batikan Koroglu, LLNL, USA	Improvement of drop-hammer impact testing for safety assessment of aged high explosives
Harvey J. Newman, Lisette R. Warren, Colin R. Pulham, and Carole A. Morrison, University of Edinburgh, UK	Towards finding lead-free ballistic modifiers in double-base propellant
Samuel Darcy-Page, Cranfield University, UK	An investigation on the effects of multi-layer charge lines on EFP formation and performance

Fig. 4 Details of the IEC-2026.

5 IEC-2026

The 3rd IEC conference will take place at the New Gate hotel in Cardiff in Wales from 16th to 18th June 2026. More details can be found in Fig. 4.

On behalf of the IEC-2024 Steering Committee and the IEC-2024 Editorial Committee we would like to thank you for all your support and look forward to seeing you at IEC-2026 in Cardiff in June 2026.

CONTENTS

Section 1
Synthesis, Characterisation and Diagnostics

Synthesis, characterization, and energetic performance of a new copper complex based on 3,4,5-trinitro-1*H*-pyrazole

Ashfaq Afsar,*[a] Xiaojiao Liu,[b] Patrick McMaster,[c] Carole A. Morrison[a] and Colin R. Pulham[a]

The design of both initiators and pyrotechnics that are more environmentally benign ("green") relies on the use of nitrogen- or oxygen-rich materials that act as efficient oxidizing agents to reduce formation of particulates, together with less toxic metals such as copper. Herein, we report the synthesis and characterization of a new copper complex based on the trinitropyrazolate (TNP^-) ligand – ammonium copper(tetrakis)trinitropyrazolate $[(NH_4^+)_2Cu^{II}(TNP^-)_4]$, which displays remarkable thermal stability, combined with a high energy output. It is therefore a promising candidate to replace particular lead-based initiators, as well as being of interest as a pyrotechnic.

1 Introduction

Energetic materials (explosives, propellants, and pyrotechnics) play important roles in both military and civilian applications. Their sensitivities to initiation by impact, shock, friction, and spark are crucial not only for their performance, but also for safety in manufacture, processing, storage, transport, and operational use.[1] Of increasing concern is the release of harmful compounds into the environment such as carbonaceous particulates (soot and smoke), chlorine-containing compounds (perchlorates and hydrogen chloride), and heavy metals such as lead and barium.[2,3] Lead-based compounds such as lead azide and lead styphnate are still widely used as initiators in a wide range of applications (*e.g.* detonators and cartridges) and so there is a strong drive to replace them with much less toxic compounds based on metals such as copper.[4] A number of alternatives with fewer health and environmental issues have been investigated and include a family of energetic copper nitrotetrazolates consisting of $Cu(en)_2(NT)_2$, $Cu(NH_3)_3(NT)_2$, copper(I) nitrotetrazolate or DBX-1, $Cu(NT)_2$ and $Cu(H_3O)(NT)_3{\cdot}3H_2O$ (en = ethylene diamine, NT = 5-nitrotetrazolate).[4] Many pyrotechnics that are used for safety and signalling applications (*e.g.* emergency flares) contain barium compounds (*e.g.* $BaCO_3$ and $BaCl_2$) to generate vibrant green flames. However, barium salts are toxic and ingestion can cause profound hypokalemia (low-potassium levels), leading to ventricular dysrhythmias.[5] For this reason, copper(II) compounds are being explored as potential substitutes to replace toxic barium salts in pyrotechnics.

[a] *EaStCHEM School of Chemistry, University of Edinburgh, Joseph Black Building, David Brewster Road, Edinburgh, UK. E-mail: aafsar@ed.ac.uk*
[b] *Harwell Campus, Diamond Light Source Limited, Oxfordshire, UK*
[c] *MOD Abbey Wood, Bristol, UK*

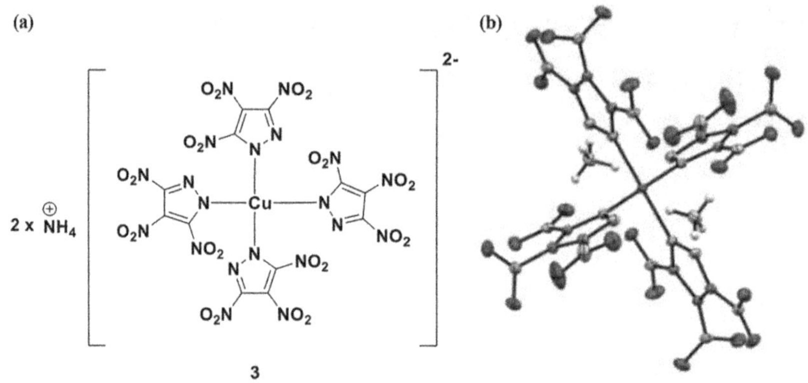

Fig. 1 (a) Molecular structure and (b) crystal structure of the complex $[(NH_4^+)_2Cu^{II}(TNP^-)_4]$.

The design of both initiators and pyrotechnics that are more environmentally benign ("green") relies on the use of nitrogen- or oxygen-rich materials that act as efficient oxidizing agents to reduce formation of particulates, together with less toxic metals. We have recently prepared and characterized a new copper compound based on the trinitropyrazolate (TNP^-) ligand – ammonium copper (tetrakis)trinitropyrazolate $[(NH_4^+)_2Cu^{II}(TNP^-)_4]$ – see Fig. 1. This compound displays remarkable thermal stability (decomposition onset at 308 °C), combined with a high energy output. It is therefore a promising candidate to replace particular lead-based initiators, as well as being of interest as a pyrotechnic.

2 Experimental

2.1 Materials and methods

Reagents were purchased from Merck and Fluorochem as analytical grade and were used as received. ^1H and ^{13}C spectra were recorded on a Bruker AVA 500 nuclear magnetic resonance spectrometer (500.12, and 125.75 MHz, respectively) with internal standard (^1H NMR: DMSO at 2.50 ppm; ^{13}C NMR: DMSO at 39.52 ppm).

A suitable single crystal of $[(NH_4^+)_2Cu^{II}(TNP^-)_4]$ was selected and mounted on a Rigaku Oxford Diffraction SuperNova diffractometer (Cu-K_α radiation, $\lambda = 1.5418$ Å). The crystal was maintained at 120 K during data collection using an Oxford Cryostream Plus. The CrysAlisPro Program was used for data processing and reduction. Structure determination was achieved using Olex2 software,[6] using the ShelXT program[7] with Intrinsic Phasing methods. Structural refinement was performed within the ShelXL program using least squares minimization.

PXRD data were collected using a Bruker D2 Phase Lab X-ray diffractometer (Cu $K_{\alpha 1} = 1.54056$ Å) operating in flat-plate geometry and the data were analysed using TOPAS.[8]

DSC measurements were performed on a TA DSC2500 over the temperature range 25–400 °C at a scan rate of 10 K min^{-1}, with sample masses of typically 1–2 mg in a T_{zero} aluminium pan. Decomposition was run in a

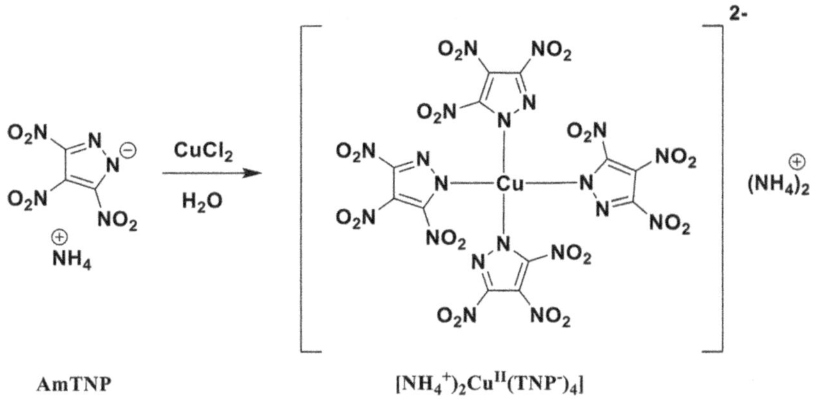

Scheme 1 Synthetic route for $[(NH_4^+)_2Cu^{II}(TNP^-)_4]$.

constant-volume high-pressure rated pan, to allow decomposition without damaging the instrument.

2.2 Synthesis of $[(NH_4^+)_2Cu(TNP^-)_4]$

$[(NH_4^+)_2Cu^{II}(TNP^-)_4]$ was prepared according to Scheme 1. The precursor, ammonium trinitropyrazole (AmTNP), was synthesised according to previously reported procedures.[9,10]

Safety warning: TNP and its salts and complexes are highly energetic and may explode in response to certain external stimuli. It is strongly recommended to carry out small-scale reactions and take appropriate precautions such as the use of leather gloves, protective coats, face shields, and explosion-proof baffles.

$CuCl_2$ (0.102 g, 0.76 mmol) was dissolved in DI water (7 mL) by heating and stirring at 70 °C. AmTNP (0.334 g, 1.5 mmol) was separately dissolved in DI water (7 mL) and heated to 70 °C. The AmTNP solution was then slowly added to the $CuCl_2$ solution, which then turned dark green. Heating and stirring was maintained for 1 hour before being allowed to cool to room temperature. The solution was then allowed to evaporate slowly overnight to give purple crystals (0.300 g, 43%). ^{1}H NMR (500 MHz, DMSO-d_6) δ 7.07 (t, $J=50$ Hz, 8H) ppm; ^{13}C NMR (126 MHz, DMSO-d_6) δ 145, 121.0 ppm.

3 Results and discussions

The crystal structure of $[(NH_4^+)_2Cu^{II}(TNP^-)_4]$ was solved and refined in the triclinic space group $P\bar{1}$ ($a=9.1943$ Å, $b=9.8905$ Å, $c=10.1633$ Å, $\alpha=72.716°$, $\beta=68.867°$, $\gamma=62.428°$). Other relevant data and parameters of the X-ray measurements are shown in Table 1.

The asymmetric unit contains one NH_4^+ cation, two TNP^- ligands and a half of a Cu(II) cation. The Cu(II) cation is coordinated to four TNP^- anions through one of the N atoms of each pyrazole ring to give a square-planar geometry. The Cu–N distances are very similar and lie in the range 2.002–2.004 Å.

Table 1 Crystallographic data and structure refinement parameters for $[(NH_4^+)_2Cu^{II}(TNP^-)_4]$.

Parameter	$[(NH_4^+)_2Cu^{II}(TNP^-)_4]$
Molecular formula	$C_{12}H_8N_{22}O_{24}Cu$
Molecular weight/g mol^{-1}	907.94
Temperature/K	120(10)
Crystal system	Triclinic
Space group	$P\bar{1}$
a/(Å)	9.1943(9)
b/(Å)	9.8905(8)
c/(Å)	10.1633(6)
α/°	$\alpha = 72.716(7)$
β/°	$\beta = 68.867(7)$
γ/°	$\gamma = 62.428(9)$
Volume/Å^3	754.61(13)
Z	1
Z'	1
Density/g cm^{-3}	1.998
F(000)	455
Radiation type	Cu
μ/mm^{-1}	2.340
θ range/°	4.6630–75.5840
Reflections collected	7187
Unique reflections	4166
Reflections $I > 2\sigma(I)$	2767
R_{int}	0.0393
Goodness-of-fit (F^2)	1.035
$R1$ ($I > 2\sigma(I)$)	0.0341
w$R2$ ($I > 2\sigma(I)$)	0.0892
CCDC	2375273

It is further stabilised by two strong intermolecular hydrogen bonds (H-bond), forming between the second N atom on the pyrazole ring and the NH_4^+ cation, with H-bond lengths of 2.921 Å and 2.933 Å, respectively. The inversion centre is located on the Cu cation, leading to a *tetrakis* formation of charge-neutral ammonium copper trinitropyrazolate (Fig. 2a). The remaining two hydrogen atoms on NH_4^+ form H-bonds with the O atoms in the nitro group on the two neighbouring $[NH_4]_2Cu^{II}(TNP)_4$ units, with bond lengths of 2.919 Å and 3.017 Å, respectively (Fig. 2b). $[Cu^{II}(TNP)_4]^{2-}$ segments form crystallographic planes in line with the 0k0 planes with layers NH_4^+ cations inserting in between (Fig. 2c). The extended 3D layer packing structure is stabilised by the strong hydrogen networks formed among NH_4^+ cations and adjacent $[Cu^{II}(TNP)_4]^{2-}$ segments (Fig. 2b).

Polycrystalline samples of $[(NH_4^+)_2Cu^{II}(TNP^-)_4]$ complex were also characterized by X-ray powder diffraction. Pawley fits were performed on the powder pattern and the pattern was fitted to the unit-cell parameters (refined to take into the account of thermal expansion) obtained from the single crystal structure using TOPAS. The PXRD pattern of the sample is shown in Fig. 3. All peaks could be fitted, thereby confirming that the single crystal is representative of the bulk powder, *i.e.* there are no additional crystalline phases or impurities in the bulk powder.

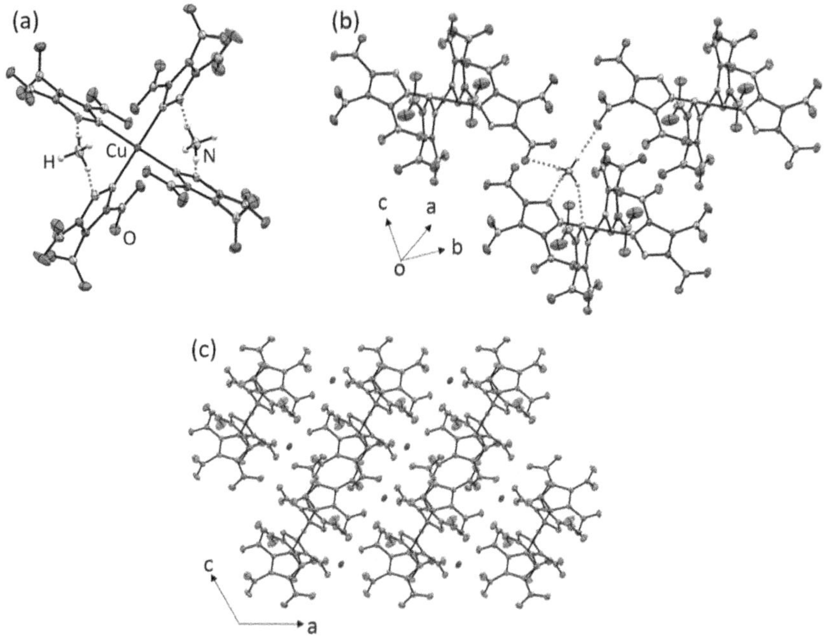

Fig. 2 (a) Crystal structure of $[(NH_4^+)_2Cu^{II}(TNP^-)_4]$; (b) the intermolecular hydrogen-bonding network with hydrogen bonds highlighted by blue dashed lines; (c) the crystal packing of $[(NH_4^+)_2Cu^{II}(TNP^-)_4]$ viewed down the crystallographic *b*-axis – ammonium cations are shown in blue, and $[Cu^{II}(TNP)_4]^{2-}$ anions are shown in red (hydrogen atoms have been omitted for clarity).

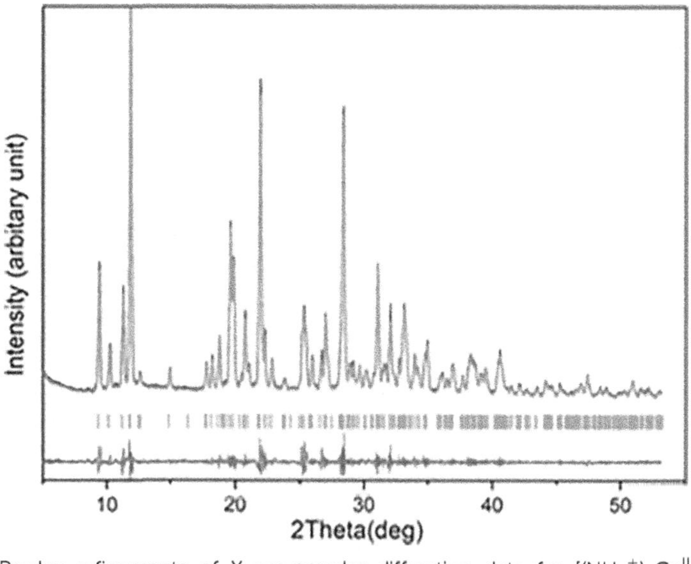

Fig. 3 Pawley refinements of X-ray powder diffraction data for $[(NH_4^+)_2Cu^{II}(TNP^-)_4]$. Experimental profile (black), calculated profile (red) and difference (blue). Vertical tick marks show refined peak positions (pink). Data were collected at 25 °C with $\lambda = 1.5418$ Å.

The DSC plot shown in Fig. 4 highlights that the compound displayed remarkable thermal stability and decomposes without melting (onset of decomposition at 308 °C) under the experimental conditions. Moreover,

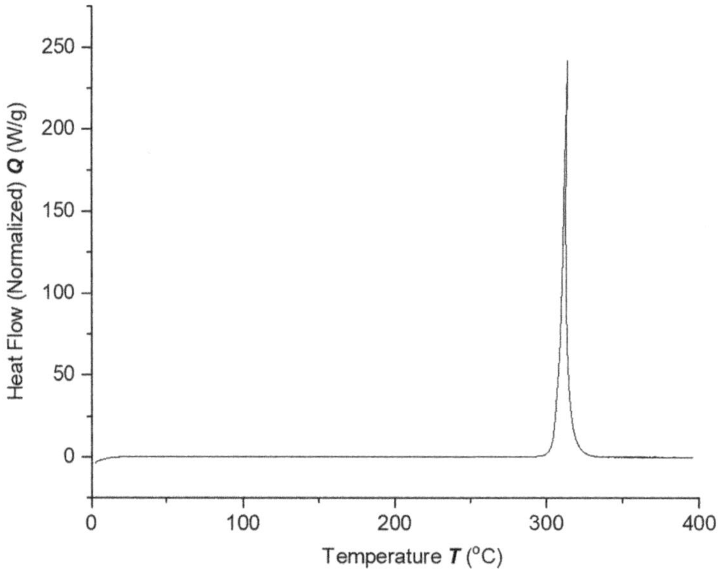

Fig. 4 DSC curve for $[(NH_4^+)_2Cu^{II}(TNP^-)_4]$.

decomposition is accompanied by a high energy output (~6 $kJ\,g^{-1}$) and a vibrant green flame. At ambient temperature, the sample was chemically stable and showed no signs of hygroscopicity even at high relative humidity (90%).

In the ^1H NMR spectrum, the hydrogen signals of the ammonium cations were observed and easily assigned at 7.07 ppm, since there was no proton associated with the TNP anion. In the ^{13}C NMR spectrum, two low-intensity signals at 121 (C4) and 145 (C3 and C5) ppm were observed assigned to the TNP anion.

4 Conclusions

In summary, we have prepared and characterized a new copper compound based on the trinitropyrazolate (TNP^-) ligand – ammonium copper(II) (tetrakis)trinitropyrazolate $[(NH_4^+)_2Cu^{II}(TNP^-)_4]$, which displays remarkable thermal stability, combined with a high energy output. It is therefore a promising candidate to replace particular lead-based initiators, as well as being of interest as a pyrotechnic. As part of a more comprehensive study of this material, we are currently exploring its behaviour under a range of physical conditions in order to assess its applicability for wider use, including sensitivity to impact, friction, and spark. This is particularly important as such materials may be subjected to a range of conditions (*e.g.* impact, friction, temperature) during manufacture, processing, long-term storage, and operation.

References

1 J. Akhavan, *The Chemistry of Explosives*, RSC, Cambridge, 3rd edn, 2011.
2 G. Steinhauser, J. H. Sterba, M. Foster, F. Grass and M. Bichler, *Atmos. Environ.*, 2008, **42**(37), 8616–8622.

3 J. W. Fronabarger, M. D. Williams, W. B. Sanborn, J. G. Bragg, D. A. Parrish and M. Bichay, *Propellants, Explos., Pyrotech.*, 2011, **36**, 541–550.
4 B. Westwater, H. J. Lloyd, I. J. Vitorica-Yrezabal, A. Fong, P. McMaster, M. Sloan, B. M. Coaker, C. R. Pulham and P. Portius, *Dalton Trans.*, 2020, **49**, 14975–14984.
5 R. Shankle and J. Keane, *Arch. Neurol.*, 1988, **45**, 579–580.
6 O. V. Dolomanov, L. J. Bourhis, R. J. Gildea, J. A. K. Howard and H. Puschmann, *J. Appl. Crystallogr.*, 2009, **42**, 339–341.
7 G. M. Sheldric, *Acta Crystallogr., Sect. A: Found. Adv.*, 2015, **71**, 3–8.
8 A. Coelho, *TOPAS-Academic* V6, Coelho Software, Brisbane, Australia, 2015.
9 Y. Q. Zhang, Y. Guo, Y. H. Joo, D. A. Parrish and J. M. Shreeve, *Chem. – Eur. J.*, 2010, **16**, 10778–10784.
10 I. L. Dalinger, I. A. Vatsadze, T. K. Shkineva, G. P. Popova and S. A. Shevelev, *Mendeleev Commun.*, 2010, **20**, 253–254.

The growth and shock testing of large single crystal energetics and analogues†

Alice Mintoff,* Colleen Williams, Chris Stennett and Christine McGee

Crystalline materials often contain mechanical and physical (and to some extent chemical) defects that may influence the intrinsic properties of the bulk material. Energetic materials (EMs) are an example of such organic crystals, and the structural defects can therefore influence behaviours, such as initiation, DDT and inert shock response. A method has been developed for growing partially defective organic crystal samples of HMX of size 10–20 mm, and this was approached by studying the growth of non-energetic organic single crystals, primarily paracetamol. Paracetamol is a good analogue for HMX, as it also exhibits crystal twinning and has similar growth conditions. Furthermore, we present a technique to understand the morphology and density of defects in both inert and energetic crystals, using a high-resolution X-ray CT scanner. This paper explores work on single crystal growth, analysis of these samples and how to then prepare those for shock testing.

1 Introduction

The use of energetic materials (EMs) is widespread within the defence industry, with a variety of combinations of energetics and polymers used in a multitude of munitions.[1] The combinations of inert and EMs are tested to ensure compositions of the materials can be safe to handle whilst maintaining the desired function; this is done in their powder form. This leads to the question as to whether the crystal structure of the energetic materials plays a part in this.

Many solid energetic materials are crystalline in nature, meaning they have a long range, repeating structure that can be simplified into one small, repeating building block known as a unit cell. Crystalline materials are used in nearly every aspect of life, from medicines to technology in most homes. Characterising both chemical and physical properties allows researchers to define and control the properties and tuneability of these. Knowledge of crystal structures allows for improvement in currently energetic materials and their compositions, through altering manufacturing processes or adding additional atoms into the materials. Knowledge of crystal structures allows for the improvement of existing energetic materials. This can be done through doping with other elements or through altering manufacturing methods. Some materials exhibit polymorphism, energetics are no exception. Octahydro-1,3,5,7-tetranitro-1,3,5,7-tetrazocine (HMX) has 4 polymorphs, each requires different manufacturing conditions and varying stability.

The crystal structure dictates the shape of the crystals, energetic materials that form needle-like crystals are more sensitive to initiation, due to higher surface area and undesirable packing.[2] The sensitiveness of

Cranfield University, UK. E-mail: alice.k.mintoff@Cranfield.ac.uk

most known energetic compositions is well established through standardised testing techniques. Although the standards differ between nations, most tests are performed on materials in the powder form and, unless performed in a clean room, there is a high chance of contamination from dust and grit. Contaminants can alter both the sensitivity and sensitiveness.[3] Sensitiveness is defined as the ease of initiation when insulted unintentionally, whilst sensitivity is the ease of initiation at the intended time. It should be noted that the sensitiveness and sensitivity is driven by the crystallinity of the component materials in the composition, due to mesostructural [10–100 μm] and microstructural [1–10 μm] defects. Defects are defined by the imperfections in the regular geometric arrangement of atoms within a crystalline solid. Growing large single crystals of energetic materials removes any impurities, which allows for a clearer picture of their behaviour and therefore how the presence of defects can alter the initiation of materials. It is possible to tune the level of defects within a material, by altering the growth method or adding in extra materials. Using powders to grow single crystals removes unintended impurities. Reducing the level of defects within a single crystal is a possibility by extending the growth period with slower cooling rates, or using a larger temperature window, but there is a trade-off between time and degree of crystal perfection.[4] Crystal growth can be achieved through different methods, and the ideal method depends on the materials used and the desired outcome. The method of crystal growth used for this line of work is the temperature lowering method, which utilises slow cooling to give a greater control of crystal growth.[5]

The central focus of this project surrounds HMX and the effect that defects have on initiation. HMX is a high explosive and is the energetic component of some commonly used plastic bonded explosives (PBX), common munitions with a largely military application.[6] It has a monoclinic crystal system with a $P2_1/n$ space group. It exhibits crystal twinning, a phenomenon where two faces of the same material share crystal lattice points.[7,8] Use of inert simulants in energetic testing is not a new concept and the common inert simulant used for HMX is 2,4,6-triamino-1,3,5-triazine, or melamine. Melamine has similar mechanical properties to HMX, so it can be used to mimic how HMX would behave when mixing to produce various PBXs. It also shares the same crystal system and space group as HMX. Like HMX, it is possible to grow melamine into a single crystal, but the ideal solvent for this isn't acetone, it's water and so has been ruled out for this testing.[9] It is difficult for each inert to share all the same properties we see in HMX and as such, each inert will be specifically designed to mimic only one or two specific formulations. Acetaminophen (paracetamol) has been used as an alternative inert simulant for HMX since the later 1900s. Much like melamine, paracetamol also has the same crystal system and space grouping as HMX, but unlike melamine, paracetamol is soluble in acetone making it a more suitable substitute for this work.[10]

Single crystals of both paracetamol and HMX were obtained through the slow cooling growth method and then analysed, using CT scanning to determine the density of void defects present. The ultimate aim of this work is to look at what effect crystalline defects have on the sensitivity of HMX and to explore this, the crystals need to be exposed to shock. Shock waves can be introduced into the sample through the use of a high-speed gas gun which allows for precise control over both speed and amplitude of the waves. Successful use of a high-speed, single stage gas gun has been used to investigate a range of materials and their mechanical properties.[11]

The following paper explores how shock waves travel through single crystals of paracetamol, as a proof of concept for studying how shock waves interact with defects present in single crystal nitramine explosives.

2 Experimental

2.1 Crystal growth

All experiments were carried out using a thermo-scientific temperature-controlled water bath. The solubility for both HMX and paracetamol was tested by dissolving an excess of solid into acetone (99.8%). The system was sealed and left to stir (25 rpm, overhead stirrer) at a defined temperature for 24 hours. The mixture was then left to stand at temperature without stirring for 1 h, after which three aliquots (3 mL) were taken.

Paracetamol was obtained from Merck and purified *via* recrystallisation. Paracetamol (purified, 11.16 g) was dissolved in acetone (99.8%, 142 mL) then filtered under vacuum. The filtrate was transferred to a crystallisation dish and partially covered to allow for crystallisation *via* evaporation for 7 days. This yielded small crystals of between 5 and 10 mm. These were placed into a second crystallisation dish with additional supersaturated solution to increase their size to between 10 and 15 mm.

Larger seed crystals were selected and were drilled through centrally (0.6 mm diameter), a nylon thread was passed through and fixed with an over-hand knot. The crystal was then suspended in saturated solution and stirred at 22 rpm. A growth rate of 66 $K\,h^{-1}$ was used with a starting temperature of 313 K and finishing temperature of 293 K. The same procedure was used to grow both seeds and large single crystals of HMX.

HMX was obtained in-house and was purified *via* recrystallisation. HMX (purified, 1.8 g) was dissolved in acetone (99.8%, 150 mL) then filtered under vacuum. The filtrate was transferred to a crystallisation dish and partially covered to allow for crystallisation *via* evaporation for 14 days. The seeds from this varied in size measuring between 2 and 7 mm and were put through a second crystallisation process with an additional supersaturated solution to increase size for ease of handling.

Seed crystals between 7 and 10 mm in size were attached to nylon thread using an isocyanate-cured polyurethane two-part adhesive with a 24-h cure rate. Other glues with appropriate compatibility were considered and tested, but the addition of acetone resulted in the glue peeling away in solution. The use of glue for HMX was necessary, as mechanically drilling a hole into the HMX seeds results in secondary nucleation points around this space. The slow-cooling method was the chosen growth method of single crystals. This method gives greater control over the growth rate and therefore development of defects.

A cooling rate of [0.05 $K h^{-1}$] was used with a starting temperature of 313 K lowering to 293 K.

Faces of the finished single crystals were then indexed using a protractor and stereographic projections.

2.2 Target specimen preparation and characterisation

To obtain slices for the plate impact experiments, the single crystals were set into paraffin wax and then sliced using a rotary saw with a diamond edge blade. The saw clamp requires samples that are uniform on all sides, the wax allowed for this, with an additional wax bottom to allow for even slicing through the material.

A schematic diagram of the shock experiment arrangement is shown in Fig. 1. The specimen slices were mounted onto a mounting plate. Each mounting plate had 3 paracetamol slices, using an isocyanate glue and left to cure for 24 hours before CT scanning for further analysis of heights. A perspex impedance window was used for 3 of the 12 shots, as this material was found to be the best match of impedance to paracetamol. Fig. 2 details the finished target plate, detailing the placement of the samples on the left and the mechanism used to affix the heterodyne velocimetry channel to measure the speeds of the wave.

2.3 Shock response and Hugoniot determination

A single-stage gas gun, with a 6 m long barrel, and 50 mm calibre was used to produce planar one-dimensional shock waves. This method is described by the work Bourne and Stevens undertook in 2001.[11] The impact experiments took place at 3 different speeds that ranged from ~400 to ~700 $m s^{-1}$, to produce different input shock pressures. These were achieved using air as a carrier gas at pressures of 50 and 150-bar and helium as a carrier gas at a pressure of 150-bar.

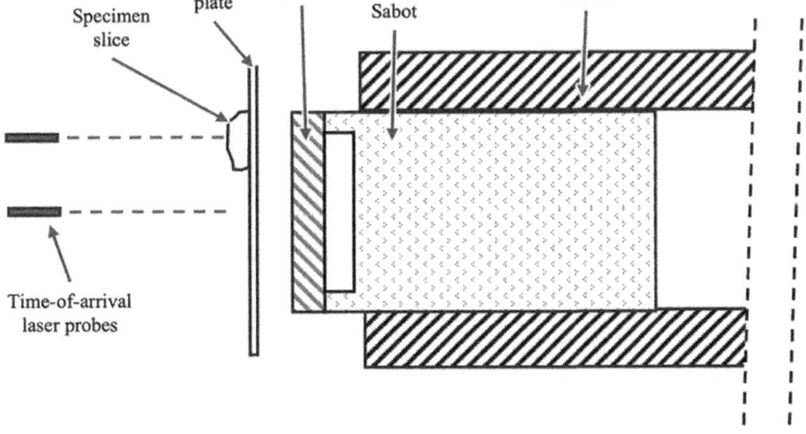

Fig. 1 Schematic of shock experiment. Flyer disk travels from right to left, impacting the specimen mounting plate.

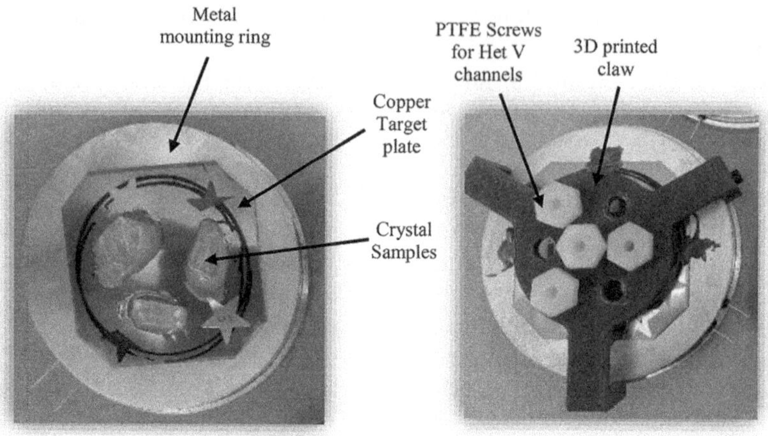

Fig. 2 Target set up for impact experiments.

For each specimen, the shock velocity was obtained by dividing the thickness of the specimen (measured from CT analysis, as described above) by the time difference between entry and exit of the shock wave through the specimen slice at the front and rear faces. The arrival of the shock at the front face of the specimen was assumed to be simultaneous with the emergence of the shock from the copper mounting plate, as observed by the central interferometer probe. The arrival of the shock at the rear face of each specimen was determined by additional interferometer channels.

Two methods were attempted to determine the particle velocity associated with each shock. In the experiments that employed PMMA impedance windows, it would be expected that the motion of the interface between the specimen and impedance window would follow a similar 'rectangular' wave profile as the input shock. From this, it would be possible to determine the particle velocity by the well-known impedance matching technique. In those experiments without impedance windows, the motion of the free surface at the rear face would be expected to show a velocity equal to twice the particle velocity.

3 Results and discussions

The growth of crystalline materials is down to both the kinetics and the thermodynamics of the molecules interacting. The interaction between molecules is down to the intermolecular forces and how these dictate the way molecules interact with each other. These interactions occur in a range of configurations due to different thermodynamics, but the configurations most likely to be seen are the more thermodynamically favourable ones. This is often what leads to polymorphism, some molecules have more than one ideal thermodynamic state.

Crystals with a longer growth period are bigger. There is more time for the solute to come out of solution onto the starting seed. Heating up the sample in an appropriate solution results in both molecules of the desired material and the impurities falling out of solution at different times so that material can grow as a single crystal without the addition of impurities. If the crystal geometry of the material in solution matches that of the seed crystal, it is more likely that the material will remain on the single crystal rather than going back into solution. Slow growth results in ideal growth conditions, but there is a trade-off between growth time and the quality of crystal. Paracetamol crystals grown over 24 hours weren't seen to have many more defects than those grown over 72 hours, but there was a significant increase in the number of cleavage planes which made the crystals unsuitable for obtaining samples.

There are many defect types that can be present in any given crystal structure; these have the ability to change material properties which depending on application, can be advantageous. For the purpose of this study, it is only possible to identify void defects – the absence of material – which are expected to be solvent inclusions, but due to the sample preparation method, it is not possible to say for certain which are solvent filled voids and which are air.

3.1 Crystal characterisation

The crystal defects found in the crystals used within this study were found to contain defects within the macrostructure. Using CT scanning, it was possible to determine the presence of these voids, although not whether the voids contain air, or part of the growth solution. It is possible the crystals did have microscopic defects, but various limitations meant that this was outside of the scope of this project. The use of CT scanning was not limited to determining the number of void defects present in the crystal samples, but also allowed for the accurate measuring of samples which is important for future shock impact work. The different growing rates used to form both the paracetamol and HMX single crystals resulted in different densities of void defects within the crystals. Crystals grown at a faster rate resulted in a higher percentage of defects, as seen in Table 1.

Initial CT scanning was performed on the crystal as a whole and then after the crystal was sliced for shock work, further CT scans were taken on these slices. It was harder to determine the true percentage defects, as the glue used to attach the slices onto the copper plate produces air bubbles

Table 1 The different growth rates of crystal samples and the percentage of void defects they contain.

Sample	Material	Growth rate (K min^{-1})	% Defects
067	Paracetamol	0.63	3.2
081	Paracetamol	0.63	2.8
082	Paracetamol	0.63	3.7
083	Paracetamol	0.63	3.9
084	Paracetamol	0.63	2.4
075	HMX	0.63	1.2
089	HMX	0.63	0.5
090	HMX	0.63	1.3
091	HMX	0.63	1.6

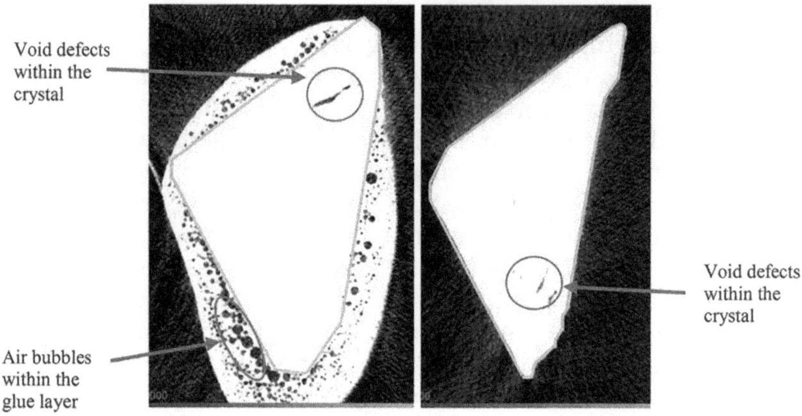

Fig. 3 Two CT images that show different slices of the same single crystal slice glued onto a copper target plate. The left-hand image is a slice at the bottom on the space next to the glue and the right-hand image is a slice towards the top of the sample.

as it cures. These additional bubbles of air do not affect the defect density in the slices of crystal, but they do potentially affect how the uniform shock wave interacts with the crystal on the other side of the glue layer. Fig. 3 shows two image slices of the same crystal sample. The left image shows a section of the crystal sample at the bottom, closer to the copper mounting plate and the right image shows a slice further away from the copper plate. The red circle denotes the air bubbles as a result of the glue drying, whilst the green circles show the void defects. The air bubbles from the glue are a round, uniform shape, whilst the voids within the crystals are not, these are longer and thinner in shape and more isolated. As the image slices move through the crystal, the round bubbles on the outer edges of the sample disappear which contributes to the idea that these are a result of the glue layer.

Another benefit of using CT scanning on the crystal samples is the ability to use the resulting images for accurate measurement of sample thickness. To carry out shock impact experiments, it was important to have a sample of known thickness that didn't have a depth greater than the length or width of the sample. Measurements can be carried out mechanically using callipers and a drop gauge which gives the required accuracy. The benefit of using CT scanning for this task is that it is possible to take these measurements after the target is assembled with all samples. This means that losses of material during target assembly are accounted for using the measurements from CT micrographs. The total volume of material can be determined and the size of the voids that may not be visible otherwise and then it is possible to know how much of the sample contains void defects.

3.2 Target preparation

Targets for single stage gas guns require a flying plate and samples of uniform thickness. Much thought was given to how the single crystal paracetamol could be sliced into smaller sections. Initial ideas were to use a thin thread that is slowly rotated and passed through an acetone solvent bath. This process proved to be slow and not reliable for getting slices with a uniform thickness. Further thought was given to using a rotary saw, and a diamond edged blade of less than 1 mm was chosen to test this method. An acetone solvent bath was used as this also gave the benefit of etching the surface of each slice, which gives a cleaner surface. Use of a rotary saw requires a regular shape and so the single crystal paracetamol were set in paraffin wax in a regular cube shape. This shape allowed for the saw clamp to grip it and meant that all of the crystal could be sliced up and the best slices were then used for target building. To ensure completely flat samples, the samples were then sanded down using wet and dry paper. Although the use of solvent soaked string did provide slices, these required sanding down and this is not suitable for the HMX crystals that this work is preparation for. The use of a solvent saw eliminates this need, making it a safer process for future work. The saw used for target preparation can also be adapted for remote operation so is suitable for use with energetic materials, as the operator can be behind a safety screen, or in another room when the saw is being used.

The glue required for attaching samples to the target plate is required to be epoxy-free so that there is no disadvantageous reaction between the glue and the HMX, so this same glue was also used for the paracetamol samples. The glue holds samples well, but as it cures, air bubbles are produced which has the potential to interfere with the planar shock waves. Current results from shock experiments are inconclusive and more experiments are required to look at whether there is any interference.

4 Conclusions

In this study, single crystals were grown of both HMX and paracetamol, with the end goal to carry out shock impact experiments on both materials to determine the correlation between crystalline defects and behaviour at impact. The scope of this paper looked at the single crystal growth and preparation of samples with paracetamol with a goal to use these methods to prepare HMX targets.

Crystal growth equipment was developed and adapted to be able to grow the largest single crystals out of solution. The crystal growth methods are effective, producing a range of both paracetamol and HMX crystals with varying void defect density. The creation of target plates for the single stage gas gun experiments has been a success and will only require minor tweaking for use with HMX. Although gas gun experiments were carried out with paracetamol, the data produced is unclear and requires further analysis. This work will form the basis for future data acquisition for not only the paracetamol shock-Hugoniot work, but also for future data collection with HMX.

Acknowledgements

We thank AWE and the UKRI for funding this project.

References

1 J. Akhavan, *The Chemistry of Explosives*, RSC Publishing, Cambridge, 3rd edn, 2011, p. 193.
2 P. C. Myint and A. L. Nichols, Thermodynamics of HMX Polymorphs and HMX/RDX Mixtures, *Ind. Eng. Chem. Res.*, 2017, **56**(1), 387–403.
3 O. Feurer, Influence of natural grit contamination and accelerated ageing on the sensitiveness of 2,4,6-trinitrotoluene (TNT) to mechanical insult, DA - Explosive Ordnance Engineering, 2022. Available at, http://dspace.barrington.cranfield.ac.uk/jspui/handle/123456789/363.
4 M. F. C. Ladd and R. A. Palmer, *Structure Determination by X-ray Crystallography*, Plenum Press, New York London, 3rd edn, 1993.
5 B. R. Pamplin, Introduction to Crystal Growth Methods, in *Crystal Growth*, ed. B. R. Pamlin, Pergamon, 2nd edn, 1980, ch. 1, pp. 1–21.
6 K. Ding, X.-J. Wang, Z.-P. Duan, Y.-Q. Wu and F.-L. Huang, Shock response of cyclotetramethylene tetranitramine (HMX) single crystal at elevated temperatures, *Def. Technol.*, 2023, **21**, 147–163.
7 W. Clegg, S. J. Teat, K. V. R. Prasad, R. I. Ristic, D. B. Sheen and E. A. Shepherd, *et al.*, An Oxidatively Coupled Dimer of Paracetamol, *Acta Crystallogr., Sect. C: Cryst. Struct. Commun.*, 1998, **54**(12), 1881–1882.
8 H. G. Gallagher, J. N. Sherwood and R. M. Vrcelj, Growth and dislocation studies of β-HMX, *Chem. Cent. J.*, 2014, **8**(1), 75.
9 E. W. Hughes, The Crystal Structure of Melamine, *J. Am. Chem. Soc.*, 1941, **63**(6), 1737–1752.
10 J. D. Yeager, A. L. Higginbotham Duque, M. Shorty, P. R. Bowden and J. A. Stull, Development of inert density mock materials for HMX, *J. Energ. Mater.*, 2018, **36**(3), 253–265.
11 N. K. Bourne and G. S. Stevens, A gas gun for plane and shear loading of inert and explosive targets, *Rev. Sci. Instrum.*, 2001, **72**(4), 2214–2218.

Calculating the heat of explosion for non-ideal explosives

Gareth Collett,* Jacqueline Akhavan, Rachael Hazael and Richard Critchley

Non-ideal explosives are typically highly porous, low-density heterogeneous materials. They differ from ideal explosives in that fuel and oxidiser are separated at the molecular level. As such, non-ideal explosives demonstrate lower detonation pressures and velocities than ideal explosives, with more extended detonation reaction zones. These longer detonation reaction zones produce higher heat of explosion due to secondary and tertiary combustion reactions occurring behind the detonation front. Non-ideal explosives are increasingly used as homemade explosives (HME), and it is often challenging to determine likely performance without detailed chemical analysis, or an understanding of loading density, critical diameter, and stoichiometry. Detailed analysis is not an option in security-related settings where preservation of life and property is often challenged by time. Using Hess' law of enthalpy, and empirical and quantitative thermochemical methods to determine the products of detonation, this paper introduces a simple method to determine the heat of explosion of any non-ideal explosive composition during three specific stages: anoxic detonation; anaerobic combustion; and aerobic combustion. The values calculated compare favourably to heats of explosion for non-ideal explosives published in open-source literature, suggesting that the approach can be developed into a user-friendly application for first responders.

1 Introduction

In the decade preceding COVID-19, non-ideal explosives were used in 43% of deliberate acts of explosive violence in Europe,[1] with the United Kingdom, France, and Spain having accounted for over 65% of such attacks.[2] Identification of explosive chemical precursors (ECP), which may be incorporated within an improvised explosive device (IED), can be hugely problematic for first responders and the public.[3] The choice of ECP available to criminals is broad, given the limited regulation applied to some energetic compounds.[4] As such, first responders across the international arena may be required to apply rapid judgement on the explosive performance of HME compositions. This is because of the overriding philosophy to preserve life, preserve property and return situations to normality within sensible timeframes.[5,6] Judgement has been based on the knowledge of more conventional homogeneous explosives, not the heterogeneous nature of criminal compositions, with evacuation distance based on a bare charge mass equivalent of TNT.[7,8]

However, many non-ideal explosive compositions, particularly those based on nitrate esters, can be much more powerful than TNT[9,10] and are relatively simple to synthesise,[11,12] suggesting that greater refinement of safety distance may be necessary.[13,14]

Whilst media attention has often focused on typical non-ideal explosive compositions comprising ammonium nitrate (AN) (Fig. 1), regulation

Cranfield University, UK. E-mail: gareth.collett@cranfield.ac.uk

has improved considerably for the most frequent HME encountered.[15] However, as supply chains become more heavily regulated, other fuels and oxidisers, or multiple mixtures of fuels and oxidisers, will be selected out of necessity and availability – this is simply the economics of supply and demand.[16] This trajectory has already been witnessed in Afghanistan, where regulation of ammonium nitrate led to the prolific use of potassium chlorate.[17]

Therefore, such compositions must be assessed quickly and effectively at the point of discovery based on available information. This information could come from intelligence provided to the first responder, on the chemical labels visible at the crime scene, on the physical appearance of partly synthesised or final product, the results of rapid chemical testing (if accessible), witness testimony, or affirmation from technical experts should they be available. The authors conclude that several '*ifs*' face a first responder in what can only be described as a particularly challenging environment when timely decision-making is required.

Zeldovich,[18] von Neumann,[19] and Doering[20] (ZND) explicitly described a one-dimensional detonation model for ideal explosives, comprising a leading shock front followed by a chemical reaction zone. For ideal explosives, all the potential energy contained within the explosive is released almost instantly in a short reaction zone,[21,22] with the typical ZND model describing the following:

- A vertical shock front (with associated ignition temperature (T_i), detonation pressure (D_p) and density (ρ_o) within a CHNO explosive undergoing detonation;
- induction/reaction zones (where the dissociated molecules of the explosive form the principal products of detonation[23]); and
- a non-driving reaction zone where the products of detonation react with other species.

In this latter non-driving reaction zone, most additives to CHNO compositions compete with the gaseous detonation products to increase the heat of explosion (Q_e) and alter the volume of gaseous products produced.[24] The heat of explosion is therefore driven by the enthalpies of formation of the explosive products ($CO_{(g)}$, $CO_{2(g)}$, $H_2O_{(g)}$, $NO_{2(g)}$ *etc.*) dependent upon explosive composition, oxygen balance and stoichiometry. There is an order of precedence as to which products are formed first.[25]

In current literature, the performance of ideal explosives comprising CHNO can be determined through the application of empirical and quantitative thermochemical methods published by Kamlet-Jacobs,[26] Kistiakowsky–Wilson, Springall-Roberts,[27] and Cooper.[21] Consideration has also been given to the performance of conventional explosives enhanced with highly reactive atoms such as chlorine, fluorine and aluminium,[28] which are therefore non-ideal explosives by definition.

In contrast to the ZND model, non-ideal explosives can be considered to have multiple reaction zones behind the detonation shock front, as shown in Fig. 2, which influence performance. The line *AB* depicts the

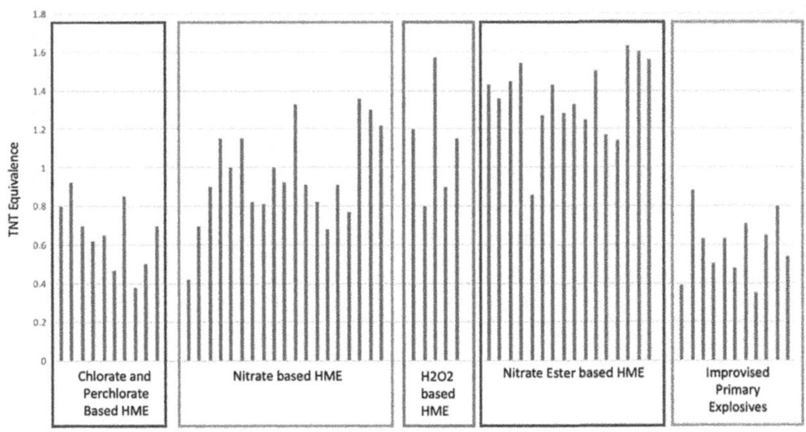

Fig. 1 TNTe of typical non-ideal explosive compositions (Q_V).[14]

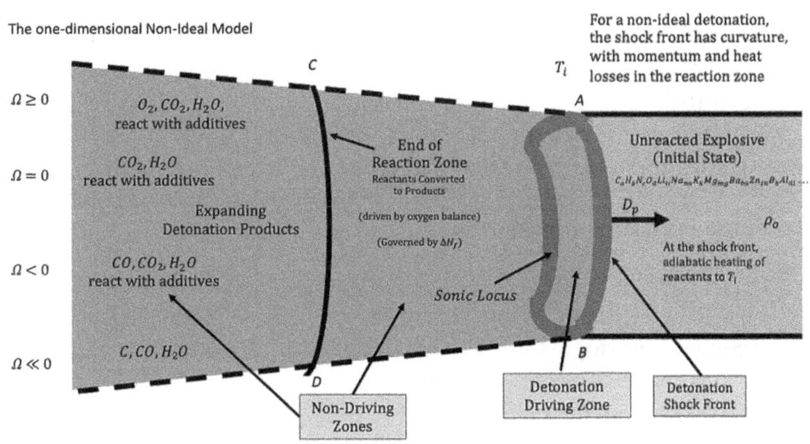

Fig. 2 Schematic of detonation reaction zones in a non-ideal explosive.

detonation shock front, with *CD* the point at which reactants are converted to products. The sonic locus depicts the region of the detonation driving zone where energy is released to sustain the shock front. Scott-Stewart and Kasimov[29] describe the sonic locus as an exact balance of heat release (Q) and flow divergence. The flow speed behind this region is sonic relative to the shock front and is therefore part of the non-driving reaction zone. As such, energy release is slower, detonation pressures lower, with a lack in oxygen balance being a contributory factor.

For example, in the case of a CHNO explosive such as TNT and RDX, the detonation velocity is typically proportional to the heat of explosion (Q), meaning that the higher the value of Q, the greater the detonation velocity.[21,23,30] However, this is not true for non-ideal explosives. Adding aluminium powder increases the heat of explosion,[31] which is dominated by mass diffusion and conduction.[32] At the same time, the detonation velocity is decreased because of the endothermic effect of the aluminium powder during the primary reaction stage.[33]

As such, the addition of aluminium generates high thermal loading[34] and produces an extended pressure pulse[35] when it reacts with the gaseous detonation products. The smaller the particle size (increasing surface area) and the nearer the approach to stoichiometry (the proportions where elements or compounds react fully with one another), the higher the heat of explosion.[36] However, if too much fuel is added, as can often be the case with HME, or the particle size of the additive is too large, this detracts significantly from the detonation velocity to the extent that detonation may not occur at all.[24,37]

As such, one would expect non-ideal explosives comprising of finely powdered fuels or metal flakes, and oxidisers with excess oxygen, to have a lower detonation velocity than their homogeneous counterparts,[29] but release significant energy upon detonation in a three-stage process outlined below (see Table 1).

The energy release in Stages 1 and 2 represent the amount of energy available for the explosive to perform mechanical work on the surroundings, whilst Stage 3 contributes to additional thermal energy that the explosive could deliver over a period of milliseconds (excess fuel and atmospheric oxygen dependent). Therefore, substance identification and an understanding of energy release are significant to the first responder, not least because of the mechanical work that the explosive can inflict on the surroundings (such as blast overpressure, impulse and fragmentation) but also because of potential thermobaric injuries caused during the aerobic combustion phase.

2 Method

This paper reviews open-source approaches used to determine the oxygen balance and heat of explosion of ideal and non-ideal explosive compositions. It has sought clarification from organisations such as Fraunhofer Institute for Chemical Technology (ICT) to determine that existing modelling tools do not establish the detonation products for multiple HME formulations.[46]

Based on the methods devised by Kistiakowsky–Wilson, Springall-Roberts, Kamlet-Jacobs, Keshavarz and Cooper,[21,25–27] this paper offers a generic chemical equation to predict the heat of explosion (Q_e) and TNT equivalence (TNTe) for any non-ideal explosive that may take the form of eqn (1):

$$C_aH_bN_cO_dLi_{li}Na_{na}K_kMg_{mg}Ba_{ba}Zn_{zn}B_bAl_{al}Pb_{pb}F_fCl_{cl}S_sSr_{sr}Br_{br}I_i \quad (1)$$

This equation considers not only current HME formulations where regulation is applied,[44] but also those formulations that may be devised as supply chains become constrained.[13,14,47] Typical compositions could, therefore, contain any number of the fuels and oxidisers.

TNTe has been chosen as the comparator, given its doctrinal use in energetic performance on land[48] and at sea[49] for over a century. However, it is acknowledged that TNTe for high and low explosives depends on the measurement method, and errors exceeding 30% are typical.[50] This is a significant margin for the first responder in terms of blast, fragmentation,

Table 1 Energy release sequence in an explosive energised by aluminium.

Stage	Event	Timescale
1	Anoxic detonation – heat is produced from the *'detonation driving zone'* as the shock front dissociates explosive molecules. No apparent temperature rise from ignition temperature (T_i) is observed until the detonation products form.[23,25,39,40] The heat of explosion (Q_e) can be calculated based on the molecular composition of the non-ideal explosive and Hess' law,[38] regardless of the physical charge structure. The explosion products *e.g.* $CO_{(g)}$, $CO_{2(g)}$, and $H_2O_{(g)}$ are first produced from the active radical species, based on thermochemistry and oxygen balance. Additives do not participate in this highly exothermic reaction zone and act as an inert material, even absorbing and consuming part of the heat liberated, thereby reducing the total energy of the detonation wave. A drop in the detonation pressure (D_{cj}/D_p) denotes this.	The induction zone can take as little as 0.1 ns to dissociate the explosive into radical species. The reaction zone for ideal explosives such as HMX and TNT ranges from 20–80 ns.
2	Anaerobic combustion – the explosion products produced in Stage 1 then react with additives under the high temperature and pressure environment behind the C–J plane. Dependent upon the combustion products, the heat of explosion increases at the expense of gas volume. For metal explosives, the major products that liberate heat other than water, carbon monoxide and carbon dioxide can be metal carbonate, metal hydride, metal hydroxide, metal nitrate, metal nitride, and metal oxide.[25,37,41]	10's to 100's of μs.
3	Aerobic combustion – if any additives remain within the non-ideal explosive after anaerobic combustion finishes, reactions then take place with oxygen in the atmosphere. It is assumed that full combustion takes place with atmospheric oxygen over a period of ms, and that this thermal energy release does not contribute to mechanical work, but rather the additional heat energy available to interact with the environment.[37,42–45]	10's of ms

safety distance, or indeed the thermal energy available to inflict quaternary physical injury.[45] For example, the values of TNTe for ideal and non-ideal explosives identified by the ballistic mortar test[51] are entirely different to that proposed by other methods such as peak overpressure/positive impulse[52] and the lead block test.[53] As such, it is essential to standardise the measurement by which any calculator operates, and base measurement upon theoretical and/or experimental literature findings. For this paper, the reference used is the heat of explosion under constant volume conditions (Q_v) where data for some non-ideal compositions in kJ kg^{-1} can be found for Stage 1 and 2 reactions.[10,54,55] A comparison of Q_e (the heat of explosion calculated for those Stage 1 and 2 reactions) is then conducted against Q_v (the reference value) to determine accuracy. This paper does not calculate the potential thermal energy from any Stage 3 reaction since it does not contribute to the mechanical work available to the explosion products.[37] TNTe for each explosive is then calculated as follows (see eqn (2)), with Q_{tnt} taken as 4091.6 kJ kg^{-1} from literature:[27]

$$\mathrm{TNT_e}\ (\mathrm{kJ\ kg^{-1}}) = \left(\frac{Q_e}{Q_{tnt}}\right) \tag{2}$$

A 4-step process is then applied to the methodology, as can be seen in Table 2, with worked examples to underpin the theory at Appendices.

3 Assumptions

Several assumptions underpin calculations given the complexity of thermochemical reactions during detonation and combustion. Table 3 outlines these assumptions to provide a simplistic baseline for the proof of concept.

Enthalpies of formation used for calculation are listed in Table 4. Unless stated otherwise, all enthalpies are provided to one decimal place. These enthalpies measure the relative thermochemical stabilities of the molecules, intimately related to their structures. For example, carbon dioxide displays greater thermal stability than carbon monoxide, and aluminium oxide displays greater thermal stability than barium oxide. Values for Q_e and Q_v are presented as kJ kg^{-1} and kJ mol^{-1}.

4 Results and discussion

4.1 Derivation of baseline programme in excel to determine oxygen balance and heat of explosion – STEP 1

The equation for a $C_aH_bN_cO_d$ explosive oxygen balance can be seen in eqn (3), which is widely published in literature.[27] The letters in subscript (a, b, c, and d) refer to the total number of atoms of carbon, hydrogen, nitrogen and oxygen in each explosive molecule. The contribution of nitrogen (c) to the oxygen balance is zero, since nitrogen does not bind to the other elements, and is therefore absent from calculation:

$$\Omega = 15.999 \times \frac{\left(d - 2a - \frac{b}{2}\right)}{\text{molecular weight}} \times 100\% \tag{3}$$

Table 2 Outline method.

STEP	Description
1. The establishment of a baseline programme in excel	As this is a proof of concept, an excel spreadsheet is constructed to determine oxygen balance (Ω) and heat of explosion for ideal compositions comprising empirical formula CHNO. The molar proportions of each atom in one gram of explosive are established to determine Q_e from Hess' law. Kistiakowsky–Wilson and modified Kistiakowsky–Wilson rules are used as the primary vehicle to develop the likely products of detonation based upon the proportion(s) of fuel(s) and oxidiser(s) and detonation theory. The spreadsheet values are then interrogated against known literature values of Q_v for those ideal explosives to determine baseline accuracy of Q_e and Ω calculations. Comparison of Q_e/Q_v should be unity.
2. Derivation of formula for non-ideal explosives	A formula is then devised for the heat of explosion for non-ideal compositions, which could take any form from eqn (1). This formula is then tested against the molecular formulae for the ideal CHNO explosive compositions examined in Step 1 to ensure that the values of Q_e and Ω obtained are identical. If this is not the case, then the formula in Step 2 is fundamentally flawed. Assumptions are clearly articulated to facilitate understanding and to influence refinement/automation of future models.
3. The introduction of powdered metals	Since so much literature is available on aluminium as an additive, aluminium is then introduced to the revised non-ideal explosive formula for TNT and ammonium nitrate (AN) compositions. This is to gain confidence in the thermochemical calculations. The oxygen balance and heat of explosion are then calculated and compared against open-source literature values using the energy release sequence in Table 1. Results of Q_e are compared to Q_v values reported in literature. Values of TNT_e are then calculated using eqn (2).
4. The introduction of varied fuel and oxidiser mixtures	Non-ideal explosives of varied molecular composition are then interrogated with Q_e/Ω compared to open-source literature values. This step determines that the calculator can manipulate explosive compositions containing multiple fuels and oxidisers with confidence.

Considering eqn (1), many fuels can be added to a CHNO explosive, such as Al, B, Ba, Ca, Cl, F, K, Li, Mg, Mn, Na, Pb, Hg, S and Zn. Each will behave in its own unique fashion. To determine the oxygen balance for such non-ideal compositions, the following principles of chemistry are considered, based on the oxidation–reduction reactions described in Table 1 and the enthalpies of formation that accompany the reactions ((ΔH_f) in kJ mol^{-1}):

1. Since C is oxidised to CO_2, each C atom requires $2O = 2a$: ΔH_f −393.5
2. Since H is oxidised to H_2O, each H requires $O/2 = b/2$: ΔH_f −241.8

Table 3 Assumptions.

Assumption	Description
1.	Atoms want to achieve their highest oxidation state, which is also their lowest internal energy state. This means that the products of detonation of $C_aH_bN_cO_d$ explosives will strive to form $CO_{2(g)}$ and $H_2O_{(g)}$, dependent upon available oxygen (Kistiakowsky–Wilson and modified Kistiakowsky–Wilson rules apply).
2.	Additives do not react with $CO_{(g)}$, $CO_{2(g)}$ or $H_2O_{(g)}$ until those products of detonation have been formed from the active radicals.
3.	That there is sufficient hydrogen in the non-ideal explosive to bind Cl, Br, I and F to HF, HCl, HI, HBr *etc.*
4.	The model assumes that more reactive elements of the non-ideal explosive compound form detonation products based on the hierarchy of the heats of reaction.[25] Additive oxidation will therefore commence with those detonation products with the highest enthalpies of formation. For example, $Al_{(s)}$ will compete with $CO_{2(g)}$ before it will progress to $H_2O_{(g)}$ and $CO_{(g)}$ to form $Al_2O_{3(s)}$. And, although $S_{(s)}$ and $Hg_{(l)}$ can have differing valence (S = 2, 4 and 6, Hg = 1, 2) the most favourable energy states such as $SO_{2(g)}$ (ΔH_f −296.1 kJ mol^{-1}) and $HgO_{(s)}$ (ΔH_f −90.7 kJ mol^{-1}) would be considered by the calculator.

Table 4 Enthalpies of formation used in calculations.[a]

Species	Phase	Empirical formula	Enthalpy of formation ΔH_f (kJ mol^{-1})
Aluminium	Solid	Al	0.0
Aluminium oxide	Solid	Al_2O_3	−1669.8
Ammonium nitrate	Solid	NH_4NO_3	−365.3
Barium	Solid	Ba	0.0
Barium nitrate	Solid	$Ba(NO_3)_2$	−988.0
Barium oxide	Solid	BaO	−558.1
Baratol	Solid	$C_{0.028}H_{0.02}N_{0.013}O_{0.026}Ba_{0.00038}$	−295.0
Carbon	Solid	C	0.0
Carbon dioxide	Gas	CO_2	−393.5
Carbon monoxide	Gas	CO	−110.5
Diesel	Liquid	$C_{12}H_{23}$	−355.9
Hydrogen	Gas	H_2	0.0
Hydrogen chloride	Gas	HCl	−92.3
Hydrogen fluoride	Gas	HF	−268.6
Hydrogen peroxide	Liquid	H_2O_2	−187.6
Lead azide	Solid	$Pb(N_2)_3$	468.6
Nitrogen	Gas	N_2	0.0
Nitroglycerine	Liquid	$C_3H_5N_3O_9$	−370.7
Paraffin wax	Liquid	$C_{25}H_{52}$	−2460.5
PETN	Solid	$C_5H_8N_4O_{12}$	−530.5
RDX	Solid	$C_3H_6N_6O_6$	+70.3
TATP	Solid	$C_9H_{18}O_6$	−151.18
TNT	Solid	$C_7H_5N_3O_6$	−62.1
3-NT	Liquid	$C_7H_7NO_2$	−31.5
Water	Gas	H_2O	−241.8
Wood meal	Solid	$C_{50}H_{72}O_{33}$	−945.8

[a] Notes for Table 4: enthalpies of formation used for calculations in this paper are taken from literature.[21,27,56,58–63] Equations refer to the phase of each compound (solid, liquid or gas). For example, where H_2O is quoted, this means $H_2O_{(g)}$, not $H_2O_{(s)}$ or $H_2O_{(l)}$.

3. Since N goes to N_2, there is no need of oxygen $= c \times 0 = 0$: $\Delta H_f = 0.0$
4. Since Al oxidises to Al_2O_3, each Al requires $3O/2 = 3Al/2$: $\Delta H_f = -1669.8$
5. Since B oxidises to B_2O_3, the same argument for Al applies 3B/2: $\Delta H_f = -1263.6$
6. Since Ca oxidises to CaO, each Ca requires one O = Ca – $\Delta H_f = -635.5$
7. Since Mg oxidises to MgO, each Mg requires one O = Mg: $\Delta H_f = -601.8$
8. Since Na oxidises to Na_2O, each Na requires 0.5O = Na/2: = −416.0
9. Since K oxidises to K_2O, each K required 0.5O = K/2: $\Delta H_f = -363.2$
10. Since Zn oxidises to ZnO, each Zn requires one O = Zn: $\Delta H_f = -348.0$
11. Since S oxidises to SO_2, each S requires 2O = 2S: $\Delta H_f = -296.1$
12. Since the oxidation number for H and F are +/1 respectively, each F requires one H, thereby reducing H_2/H_2O by 0.5 = -F/2: $\Delta H_f = -286.6$
13. Since the oxidation number for H and Cl are +/1 respectively, each Cl requires one H, thereby reducing H_2/H_2O by 0.5 = -Cl/2: $\Delta H_f = -92.3$
14. Since the oxidation numbers of H and Br are +/1 respectively, each Br requires one H, thereby reducing H_2/H_2O by 0.5 = -Br/2: $\Delta H_f = -36.2$
15. Since the oxidation numbers of H and I are +/1 respectively, each I requires one H, thereby reducing H_2/H_2O by 0.5 = -I/2: $\Delta H_f = +26.5$

This approach is consistent with thermochemical code methodologies used to predict the detonation velocity of organic, inorganic and mixed explosives, whereby the more reactive elements of the explosive compound (those with the higher enthalpies) react first, and form detonation products based on the hierarchy of the heats of reaction.[25]

The establishment of the baseline programme in excel from these principles was based on the manipulation of explosive compositions within eqn (3), using the appropriate Kistiakowsky–Wilson conditions. This step is not shown, as examples of calculations using eqn (3) are widely published.[27] The next step is to derive the oxygen balance formula for non-ideal explosives.

4.2 Derivation of non-ideal oxygen balance formula – STEP 2

From recent research at Cranfield University,[4] only certain ECPs are considered for HME manufacture, and predictive analysis can determine future trends.[64] As such, eqn (1) constitutes the principal fuel and oxidiser compositions that could be readily encountered within geographical supply chains. The products of detonation and combustion depend very much on the oxygen balance of the explosive composition and the Kistiakowsky–Wilson and modified Kistiakowsky–Wilson conditions. It is, therefore, important to understand the role of constituent atoms within the explosive molecule during anoxic detonation and anaerobic combustion. The following examples are provided to demonstrate the effect of oxygen balance within an explosive as detonation occurs, using known explosive compositions to derive the formula.

4.2.1 Example 1 – when the explosive is hugely oxygen deficient. When the amount of oxygen within the explosive molecule $C_aH_bN_cO_d$ is much less than carbon ($d \leq a$, or $O \leq C$, then $C_{(s)}$ will dominate the products of detonation, dependent upon the oxidation reactions between $CO_{(g)}$ and $H_{2(g)}$:

$$\therefore \text{ explosion products} \rightarrow \frac{c}{2}N_2 + dCO + (a-d)C + \left(\frac{b}{2}\right)H_2 \tag{4}$$

To determine if this is correct, consider the liquid explosive 3-nitrotoluene (3-NT), with molecular structure $C_7H_7NO_{2(l)}$. Using eqn (3), 3-NT has an oxygen balance of:

$$\Omega = 15.999 \times \frac{\left(d - 2a - \frac{b}{2}\right)}{\text{molecular weight}} \times 100\% \tag{5}$$

$$\therefore \Omega = 15.999 \times \frac{(2 - 14 - 3.5)}{137.14} \times 100\% \tag{6}$$

$$\therefore \Omega = -180.8\%$$

3-NT would require 8.75 additional $O_{2(g)}$ to form $CO_{2(g)}$ and $H_2O_{(g)}$, which means it is hugely oxygen deficient (−181%).[56] Using the Kistiakowsky–Wilson rules,[27] the products of detonation for an ideal explosive using eqn (4) would be as follows, which is consistent with literature:[65]

$$\therefore C_7H_7NO_{2(l)} \rightarrow \frac{1}{2}N_2 + 2CO + (7-2)C + \left(\frac{7}{2}\right)H_2 \tag{7a}$$

$$\therefore C_7H_7NO_{2(l)} \rightarrow 0.5N_{2(g)} + 2CO_{(g)} + 5C_{(s)} + 3.5H_{2(g)} \tag{7b}$$

However, if those atoms in Group 7 of the Periodic Table with a valance of −1 were introduced to 3-NT (such as F, Cl, Br, and I) then, based on the assumptions in Table 4, these atoms would naturally compete with hydrogen in the detonation reaction zone[25] and the products would become:

$$\frac{c}{2}N_2 + fHF + clHCl + brHBr + iHI + dCO + (a-d)C + \left(\frac{b - f - cl - br - i}{2}\right)H_2 \tag{8}$$

Note: *a, b, c, d, f, cl, br* and *i* denote the number of atoms present within the molecular formula for explosive $C_aH_bN_cO_dF_fCl_{cl}Br_{br}I_i$; all molecular weights from the periodic table are shown to two decimal places.

4.2.2 Example 2 – when the amount of oxygen exceeds that of carbon. When the amount of oxygen exceeds that of carbon, but the explosive remains oxygen deficient, then there will be insufficient oxygen to go beyond the formation of carbon monoxide and water. The following condition describes this relationship:

$$\mathrm{O}>\mathrm{CO} \quad \text{and} \quad \mathrm{H_2O}>\mathrm{O}\text{-}\mathrm{CO} \tag{9a}$$

$$\therefore d>a \quad \text{and} \quad \left(\frac{b-f-cl-br-i}{2}\right)>d-a \tag{9b}$$

To determine if this is correct, consider RDX, with the molecular formula $C_3H_6N_6O_{6(s)}$. The detonation products favour a less negative oxygen balance than 3-NT (−21.6%). As such, the products of detonation take the following form, which is consistent with literature and the Kistiakowsky–Wilson rules:[27]

$$\frac{c}{2}\mathrm{N_2}+a\mathrm{CO}+(d-a)\mathrm{H_2O}+\left(\frac{b}{2}-(d-a)\right)\mathrm{H_2} \tag{10a}$$

$$\therefore \frac{6}{2}\mathrm{N_2}+3\mathrm{CO}+(6-3)\mathrm{H_2O}+\left(\frac{6}{2}-(6-3)\right)\mathrm{H_2} \tag{10b}$$

$$\therefore \mathrm{C_3H_6N_6O_{6(s)}} \rightarrow 3\mathrm{N_{2(g)}}+3\mathrm{CO_{(g)}}+3\mathrm{H_2O_{(g)}} \tag{10c}$$

If those atoms in Group 7 of the Periodic Table with a valance of −1 were to be added to RDX (such as F, Cl, Br, and I), thereby creating non-ideal conditions, then these atoms would naturally compete with hydrogen and the detonation products would become:

$$\begin{aligned}&\frac{c}{2}\mathrm{N_2}+f\mathrm{HF}+cl\mathrm{HCl}+br\mathrm{HBr}+i\mathrm{HI}+a\mathrm{CO}+(d-a)\mathrm{H_2O}\\&+\left(\frac{b-f-cl-br-i}{2}-(d-a)\right)\mathrm{H_2}\end{aligned} \tag{11}$$

4.2.3 Example 3 – when the explosive oxygen balance tends towards zero. In the situation where oxygen remains more than that required to compete in the formation of carbon monoxide and the products of hydrogen (such as H_2O and HF *etc.*), but there is no excess of oxygen in the post-reaction phase, then carbon monoxide will oxidise to carbon dioxide. Such conditions are met when:

$$\mathrm{O}\geq\mathrm{CO}+\mathrm{H_2O} \quad \text{but} \quad \mathrm{O}\leq\mathrm{CO_2}+\mathrm{H_2O} \tag{12}$$

$$\therefore d\geq a+\left(\frac{b-f-cl-br-i}{2}\right) \quad \text{but} \quad d\leq 2a+\left(\frac{b-f-cl-br-i}{2}\right) \tag{13}$$

The oxygen balance becomes even less negative and tends towards zero, where the Kistiakowsky–Wilson rules dictate an ever-increasing presence of $CO_{(g)}$, $H_2O_{(g)}$ and $CO_{2(g)}$.

To determine if this is correct, consider PETN with the molecular formula $C_5H_8N_4O_{12(s)}$ (oxygen balance −10%). The products of detonation can be determined as follows, which equates well with literature:[25,27]

$$\frac{c}{2}N_2 + \left(\frac{b}{2}\right)H_2O + \left(a - \left(d - a - \frac{b}{2}\right)\right)CO + \left(d - a - \frac{b}{2}\right)CO_2 \tag{14a}$$

$$\therefore \frac{4}{2}N_2 + \left(\frac{8}{2}\right)H_2O + \left(5 - \left(12 - 5 - \frac{8}{2}\right)\right)CO + \left(12 - 5 - \frac{8}{2}\right)CO_2 \tag{14b}$$

$$\therefore C_5H_8N_4O_{12(s)} \rightarrow 2N_{2(g)} + 4H_2O_{(g)} + 2CO_{(g)} + 3CO_{2(g)} \tag{14c}$$

If those atoms in Group 7 of the Periodic Table with a valance of −1 were to be added to PETN (such as F, Cl, Br, and I), then these atoms would naturally compete with hydrogen and the detonation products would become:

$$\frac{c}{2}N_2 + fHF + clHCl + brHBr + iHI + \left(\frac{b - f - cl - br - i}{2}\right)H_2O + \left(2a - d + \frac{b - f - cl - br - i}{2}\right)CO + \left(d - a - \frac{b - f - cl - br - i}{2}\right)CO_2 \tag{15}$$

4.2.4 Example 4 – when an oxygen surplus is evident within the explosive. In a situation where there is an abundance of oxygen in the explosive, such as nitroglycerine NG ($C_3H_5N_3O_{9(l)}$), then unused oxygen remains after the production of carbon dioxide and water. This is described by the condition:

$$O > CO_2 + H_2O \tag{16a}$$

$$\therefore d > 2a + \left(\frac{b - f - cl - br - i}{2}\right) \tag{16b}$$

The products of detonation would comprise $O_{2(g)}$, $H_2O_{(g)}$ and $CO_{2(g)}$ as follows:[27]

$$\frac{c}{2}N_2 + \left(\frac{b}{2}\right)H_2O + aCO_2 + \left(\frac{d - \frac{b}{2} - 2a}{2}\right)O_2 \tag{17a}$$

$$\therefore C_3H_5N_3O_{9(l)} \rightarrow \frac{3}{2}N_2 + \left(\frac{5}{2}\right)H_2O + 3CO_2 + \left(\frac{9 - \frac{5}{2} - 6}{2}\right)O_2 \tag{17b}$$

$$\therefore C_3H_5N_3O_{9(l)} \rightarrow \frac{3}{2}N_{2(g)} + 2\frac{1}{2}H_2O_{(g)} + 3CO_{2(g)} + \left(\frac{1}{4}\right)O_{2(g)} \tag{17c}$$

If those atoms in Group 7 of the Periodic Table with a valance of −1 were to be added to NG (such as F, Cl, Br, and I), then these atoms would naturally compete with hydrogen and the detonation products would become:

$$\frac{c}{2}\mathrm{N_2} + f\mathrm{HF} + cl\mathrm{Cl} + br\mathrm{Br} + i\mathrm{I} + \left(\frac{b-f-cl-br-i}{2}\right)\mathrm{H_2O} + a\mathrm{CO_2} + \left(\frac{2d-b+f+cl+br+i}{4} - a\right)\mathrm{O_2} \quad (18)$$

Therefore, the valance of atoms in a non-ideal explosive containing some, or all the atoms annotated in eqn (1) can determine the overall products of detonation for a non-ideal explosive. The above process can be repeated for atoms within each Group of the Periodic Table, and mixtures containing atoms across multiple Groups.

4.2.5 Derivation of overall oxygen balance for a non-ideal explosive. Examples 1–4 demonstrate the impact of Group 7 atoms on oxygen balance and products of detonation for ideal explosives. A similar situation applies to Groups 1–6, as shown in Table 5.

As such, the basic oxygen balance equation for an ideal explosive (eqn (3)) can be refined according to the atomic composition of O, C, H, N, Li, Na, K, F, Cl, Mg, Ba, Zn, Al, B, Pb, Sr, Br, I as follows:

$$\therefore\ \Omega = 15.999 \times \left(\frac{d - 2a - \frac{b}{2} - \frac{\mathrm{Li}}{2} - \frac{\mathrm{Na}}{2} - \frac{\mathrm{K}}{2} + \frac{\mathrm{F}}{2} + \frac{\mathrm{Cl}}{2} + \frac{\mathrm{Br}}{2} + \frac{\mathrm{I}}{2} - \mathrm{Mg} - \mathrm{Ba} - \mathrm{Zn} - \mathrm{Sr} - \frac{3\mathrm{Al}}{2} - \frac{3\mathrm{B}}{2} - 2\mathrm{Pb}}{\text{molecular weight}}\right) \times 100. \quad (19)$$

This then provides the generic formula for oxygen balance for HME compositions as follows:

$$\therefore\ \Omega = \frac{15.999\left(d - 2a - \frac{b}{2} + \left(2\sum \mathrm{Valence}(\pm 4) + \frac{3}{2}\sum \mathrm{Valence}(\pm 3) + \sum \mathrm{Valence}(\pm 2) + \frac{1}{2}\sum \mathrm{Valence}(\pm 1) + \ldots\right)\right)}{\text{molecular weight}} \times 100 \quad (20)$$

Table 5 Group and valence entry points.

Group valence	Remarks
1 and 7 (valence ±1)	K, Na, F and Cl are highly reactive and possess valance +1/−1 respectively. This is therefore the entry point for atoms with valance +1/−1
2 and 6 (valence ±2)	Ca, Mg and O are valances +2/−2 respectively. This is therefore the entry point for atoms with valance +2/−2
3 (valence ±3)	Al and B require 3/2 oxygen (valance 3). This is therefore the entry point for metals with valence +3/−3
4 (valence ±4)	C requires two oxygen atoms to achieve their preferred energy state (valence 4). This is therefore the entry point for HME compositions containing atoms with valance +4/−4

Note: F, Cl, Br and I atoms are positive values in the equation above since they compete with the hydrogen in water, thereby releasing oxygen from the water. Atoms such as C, H, Mg, Li, Na and Al attract a negative value in the equation because they compete for oxygen within the molecule.

4.2.6 Example 5 – determining the detonation products for bisfluorodinitroethyl formal (FEFO) and PBX 9407 (94/6 RDX/Exon 461). In this example, Cl and F are introduced to the composition of a CHNO explosive. Detailed calculations can be found in Appendix A, with only the results highlighted in the main text.

Consider FEFO with the molecular formula $C_5H_6N_4O_{10}F_{2(l)}$ and molecular weight 320.12 g. As the explosive is non-ideal, theory dictates that eqn (19), not eqn (3), should be considered to determine oxygen balance. This equation calculates the oxygen balance for FEFO to be −9.99%. The following products of detonation can then be determined from Example 3 (eqn (14a)), where the conditions for a slightly negative oxygen balance are established:

$$\therefore\ C_5H_6N_4O_{10}F_{2(l)} \rightarrow 2N_{2(g)} + 2HF_{(g)} + 2H_2O_{(g)} + 2CO_{(g)} + 3CO_{2(g)} \quad (21)$$

The oxygen balance and products of detonation support reported literature values.[57]

Consider PBX 9407 with molecular formula $C_{1.41}H_{2.66}N_{2.54}O_{2.69}Cl_{0.07}F_{0.09(s)}$ and molecular weight 102.42 g.[28] Again, the explosive is non-ideal. One would therefore anticipate the oxygen balance and products of detonation to be based on eqn (14) and (19). Calculations at Appendix A realise an oxygen balance of −21.58%, with the following products of detonation determined:

$$\begin{aligned}\therefore\ C_{1.41}H_{2.66}N_{2.54}O_{2.69}Cl_{0.07}F_{0.09(s)} \rightarrow\ & 1.27N_{2(g)} + 0.09HF_{(g)} \\ & + 0.07HCl_{(g)} + 1.25H_2O_{(g)} \\ & + 1.38CO_{(g)} + 0.03CO_{2(g)}\end{aligned} \quad (22)$$

4.3 Comparison of homogeneous and binary compositions, with Q calculated from Hess' law – STEP 3

Using eqn (1) and (19), Table 6 provides the calculated values of oxygen balance (Ω) and heat of explosion (Q_e) for ideal and binary non-ideal explosives compositions. These are compared with published literature values of heat of explosion at constant volume (Q_v). AN/Al (88/12) is provided as the worked example at Appendix B.

AN/Al (88/12) is a composition with a positive oxygen balance of +6.92%.[55] As such, it should follow the process described in example 3 where a CHNO explosive would tend towards the formation of $N_{2(g)}$, $O_{2(g)}$, $H_2O_{(g)}$ and $CO_{2(g)}$. However, in this example, there is no carbon present so the products of detonation would tend towards $N_{2(g)}$, $H_2O_{(g)}$ and $Al_2O_{3(s)}$. The empirical formula by molar quantities within 1 g is therefore determined from calculation at Appendix B as:

$$H_{0.043976}N_{0.021988}O_{0.032982}Al_{0.004447(s)} \quad (23)$$

Table 6 Literature and formula values for Q and oxygen balance for ideal and non-ideal explosives.

	Literature values			Formula values		
Explosive	Q_v (kJ mol^{-1})	Ω	Ref.	Q_e (kJ mol^{-1})	Ω	$\left(\frac{Q_e}{Q_v} \times 100\%\right)$
TNT	−929.34 (−4091.63)[a]	−73.90	27	−929.30	−73.90	100.00
RDX	−1127.36 (−5075.50)[a]	−21.60	27	−1118.90	−21.60	99.25
RDX/TNT (60/40)	−1057 (−4684)[a]	−42.60	27	−1057	−42.60	100.00
NG	−1414.40 (−6228.46)[a]	+3.50	27	−1414	+3.50	99.99
TNT/Al (80/20)	−756.36 (−5275)[a]	−67.60	10	−786.60	−67.60	104.00
AN/Al (88/12)	−319.50 (−4934)[a]	+6.92	55	−325.15	+6.90	101.80

[a] Figures in (parenthesis) denote literature values for Q_v in kJ kg^{-1}.

Following the stages and assumptions described in Table 3, the aluminium competes with the detonation products formed, in this case H_2O(g), and any excess oxygen remaining after the formation of the detonation products. The heat of explosion for this reaction is then determined from Hess' law (see eqn (24)), with enthalpy values taken from Table 4.

$$\Delta H^{\theta}_{\text{reaction}} = \sum \Delta H^{\theta}_{\text{f(products)}} - \sum \Delta H^{\theta}_{\text{f(reactants)}} \tag{24}$$

The total heat of explosion produced during this process is shown in Table 7, along with the products of detonation at each stage.

From these results the calculator determines the accuracy for AN/Al (88/12) as 101.8% against the heat of explosion (Q_v) reported in literature.

4.4 Comparison of HME compositions comprising more complex mixtures, with Q calculated from Hess' law – STEP 4

Following success of the valence formula with simple compositions, more complex interactions were interrogated against open-source literature values for Q_v. The example used in the calculation is Baronal, a ternary mixture comprising barium nitrate, TNT, and Al (50/35/25). Calculations at Appendix C identify the molecular formula for this non-ideal explosive to be:

$$C_{0.01079}H_{0.0077}N_{0.00845}O_{0.02072}Ba_{0.00191}Al_{0.00556(s)} \tag{25}$$

As such, the oxygen balance for Baronal can be determined as −23.92% using eqn (19). There is, therefore, insufficient oxygen to produce $CO_{2(g)}$ and $H_2O_{(g)}$ during anoxic detonation (Stage 1). And, since barium and aluminium are inert *'passengers'* during that initial stage, they become major liberators of heat in the formation of their metal oxides during the anaerobic stage (Stage 2). Therefore, the products of detonation and heat of explosion can be determined from eqn (14) (see Table 8).

Table 9 shows the results obtained for seven further published non-ideal compositions and one lead-based composition (lead azide), which

Table 7 Reaction sequence for AN/Al (88/12).

Process	Products of detonation	Q_e (kJ kg^{-1})	Q_e (kJ mol^{-1})
Stage 1	$H_{0.043976}N_{0.021988}O_{0.032982}Al_{0.004447}$ $\rightarrow 0.010994\ N_2 + 0.021988\ H_2O + 0.005497\ O_2 + 0.004447\ Al$	−1300	
Stage 2	$0.021988\ H_2O + 0.005497\ O_2 + 0.004447\ Al$ $\rightarrow 0.002224\ Al_2O_3 + 0.022027\ H_2O + 0.002160\ O_2$	−3720	
Totals	$H_{0.043976}N_{0.021988}O_{0.032982}Al_{0.004447(s)}$ $\rightarrow 0.010994\ N_{2(g)} + 0.022027\ H_2O_{(g)} + 0.00216\ O_{2(g)} + 0.002224\ Al_2O_{3(s)}$	−5020	−325.15

$$\left(\frac{Q_e}{Q_v}\right) \times 100\% = \frac{5020}{4934} \times 100\% = 101.8\%$$

$$\left(\frac{Q_e}{Q_{tnt}}\right) \times 100\% = \frac{5020}{4091} \times 100\% = 122.7\%$$

Table 8 Reaction sequence for Baronal.[a]

Process	Products of detonation	Q_e (kJ kg^{-1})	Q_e (kJ mol^{-1})
Stage 1	$C_{0.01079}H_{0.0077}N_{0.00845}O_{0.02072}Ba_{0.00191}Al_{0.00556}$ $\rightarrow 0.004225\ N_2 + 0.00385\ H_2O + 0.00471\ CO + 0.00608\ CO_2 + 0.00191\ Ba + 0.00556\ Al$	−1837	
Stage 2	$0.0042\ N_2 + 0.00385\ H_2O + 0.00471\ CO + 0.00608\ CO_2 + 0.00191\ Ba + 0.00556\ Al$ $\rightarrow 0.0042\ N_2 + 0.00385\ H_2O + 0.00471\ CO + 0.000955\ CO_2 + 0.005125\ C + 0.00191\ BaO + 0.00278\ Al_2O_3$	−3675	
Totals	$C_{0.01079}H_{0.0077}N_{0.00845}O_{0.02072}Ba_{0.00191}Al_{0.00556(s)}$ $\rightarrow 0.004225\ N_{2(g)} + 0.005125\ C_{(s)} + 0.00385\ H_2O_{(g)} + 0.00471\ CO_{(g)} + 0.000955\ CO_{2(g)} + 0.00191\ BaO_{(s)} + 0.00278\ Al_2O_{3(s)}$	−5512	−611.17
$\left(\frac{Q_e}{Q_v}\right)$	$\left(\frac{5512}{4767}\right) \times 100\%$	114%	
$\left(\frac{Q_e}{Q_{tnt}}\right)$	$\left(\frac{5512}{4091}\right) \times 100\%$	135%	

[a] Note for Table 8: These values demonstrate an identical prediction of oxygen balance and Q_e within 14% of Q_v for baronal[54,66] (−4767 kJ kg^{-1}).

carries no oxygen within its molecular structure. A simple shading system depicts Q_e lying within 10% Q_v (none), and those lying within 20% (grey). The common name for each explosive is annotated where known.

Table 9 Comparison of non-ideal explosive heats of explosion to literature.

Expl	Literature (Q_v)	Literature Ref.	Products of detonation (to 5.S.F.)	Ω	Q_e (kJ kg^{-1})	$\left(\frac{Q_e}{Q_v}\right)$
ANFO (96/4) (Diesel $C_{12}H_{23}$)	−3950	67	$C_{0.00287}H_{0.05347}N_{0.02399}O_{0.03598} \rightarrow 0.01199\ N_{2(g)} + 0.02674\ H_2O_{(g)} + 0.00287\ CO_{2(g)} + 0.00175\ O_{2(g)}$	+5.61	−4438.5	1.12 (112%)
AN/Al (60/40)	−8866	24	$H_{0.02998}N_{0.01499}O_{0.02249}Al_{0.01483(s)} \rightarrow 0.00750\ N_{2(g)} + 0.00025\ H_2O_{(g)} + 0.01474\ H_{2(g)} + 0.00741\ Al_2O_{3(s)}$	−23.6	−9683.1	1.09 (109%)
ANFO/Al (92/5/3)	−4328	68	$C_{0.00359}H_{0.05285}N_{0.02299}O_{0.03448}Al_{0.00111(s)} \rightarrow 0.01149\ N_{2(g)} + 0.00039\ C_{(s)} + 0.02642\ H_2O_{(g)} + 0.00320\ CO_{2(g)} + 0.00056\ Al_2O_{3(s)}$	−1.25	−4304.8	0.99 (99%)
Barium nitrate/ TNT (10/90) – Baratol	−4029	69	$C_{0.02774}H_{0.01981}N_{0.01265}O_{0.02607}Ba_{0.00038(s)} \rightarrow 0.00633\ N_{2(g)} + 0.01636\ C_{(s)} + 0.00991\ H_2O_{(g)} + 0.00698\ CO_{(g)} + 0.00440\ CO_{2(g)} + 0.00038\ BaO_{(s)}$	−63.50	−4652.7	1.15 (115%)
AN/TNT/NG (70/ 25/5) Donarite	−3559	61	$C_{0.00555}H_{0.04438}N_{0.02263}O_{0.03573(s)} \rightarrow 0.01132\ N_{2(g)} + 0.02219\ H_2O_{(g)} + 0.00555\ CO_{2(g)} + 0.00122\ O_{2(g)}$	+3.91	−3773.4	1.06 (106%)
AN/TNT/AL (40/ 40/20) – Minol 2	−6782	66	$C_{0.01233}H_{0.02880}N_{0.01528}O_{0.02556}Al_{0.00741(s)} \rightarrow 0.00764\ N_{2(g)} + 0.00117\ C_{(s)} + 0.00330\ H_2O_{(g)} + 0.01116\ CO_{(g)} + 0.00371\ Al_2O_{3(s)} + 0.01112\ H_{2(g)}$	−39.38	−6900.1	1.02 (102%)
$KClO_3$/vaseline (90/10) – Cheddite	−4961	69	$C_{0.00572}H_{0.01190}O_{0.02203}K_{0.00734}Cl_{0.00734(s)} \rightarrow 0.00155\ KClO_{3(s)} + 0.00595\ H_2O_{(g)} + 0.00572\ CO_{2(g)} + 0.00580\ KCl_{(s)}$	+7.42	−5913.9	1.19 (119%)
Lead azide (PbN_6)	−1610	27 and 63	$Pb_{0.00172}N_{0.0206(s)} \rightarrow 0.00172\ Pb_{(s)} + 0.01030\ N_{2(g)}$	−5.49	−1609.0	1.00 (100%)

Table 10 Performance of non-ideal compositions without literature reference.

Explosive	Products of detonation (to 5.S.F.)	Ω	Q_e (kJ kg^{-1})	$\left(\frac{Q_e}{Q_{TNT}}\right)$
H_2O_2/Sawdust (80/20) Concentrated hydrogen peroxide (CHP)	$C_{0.00833}H_{0.05903}O_{0.05253} \rightarrow 0.02951\ H_2O_{(g)} + 0.00833\ CO_{2(g)} + 0.00318\ O_{2(g)}$	+10.19	−5842.9	1.43
Sodium chlorate/ nitrobenzene (72/28) (CO-OP explosive)	$C_{0.01365}H_{0.01137}N_{0.00227}O_{0.02484}Na_{0.00676}Cl_{0.00676(s)} \rightarrow 0.00114N_{2(g)} + 0.00814CO_{(s)} + 0.00569H_2O_{(g)} + 0.00551CO_{2(g)} + 0.00676NaCl_{(s)}$	−13.01	−4904.1	1.12
AN/TATP/Al (85/5/10)	$C_{0.00203}H_{0.04653}N_{0.02124}O_{0.03321}Al_{0.00371(s)} \rightarrow 0.01062\ N_{2(g)} + 0.02326\ H_2O_{(g)} + 0.00185\ Al_2O_{3(s)} + 0.00203\ CO_{2(g)}$	+0.54	−4057.0	1.00

Table 10 provides further calculation relating to 3 non-ideal compositions encountered by first responders,[1] where literature reference on performance is not readily available. Whilst these results are merely speculative, they suggest that the calculator can provide rapid determination of HME performance in a security-related setting. The common names for compositions are annotated where known.

The values of Q_e obtained from the valence formula in Table 10 present a *'worst case'* situation for the first responder and differ (in some instances) from literature due to the enthalpies of formation chosen for the more complex hydrocarbons such as fuel oil, woodmeal, vaseline and paraffin wax. These values were scarcely found, yet are necessary for the critical evaluation of data. For example, if the formula for fuel oil is taken as $C_{18}H_{38(l)}$, then the enthalpy of formation for this hydrocarbon (−378 kJ mol^{-1})[70] improves the accuracy of ANFO/Al calculations to 104%.

Additionally, barium is understood to continue energetic reactions well behind the shock front, where only a percentage of its oxygens react with carbon.[71] As such, it is reasonable to determine that experimental literature values for Baronal and Baratol may always be lower than those calculated in these examples since the equation of state considers optimum conditions.

Loading density, critical diameter, porosity, particle size, confinement and homogeneity of oxidiser/fuel mix, are also of huge importance in any effective calculation of explosive performance, since each aspect plays a role in the deflagration to detonation (DDT) transition.[29] For example, if density and critical diameter are below certain values for any explosive, then DDT cannot progress.[72]

However, some explosive compositions can behave in a unique fashion. For example, TNT compositions may experience decreasing critical diameters with increasing density (and increasing detonation velocity with loading density), whilst others such as ammonium nitrate may exhibit the opposite effects.[73] Therefore, focusing on the thermochemistry within this model mitigates significant 'unknowns' associated with HME manufacture, thereby calculating the worst-case conditions for the first responder. It removes the complexity of Chapman–Jouget detonation pressures and simplifies the thermochemistry behind driving and non-driving detonation reaction zones. Indeed, if the complexity of detonation velocity (D) were required, then perhaps it could be predicted from the Q_e values obtained in Stages 1 and 2 of the detonation process, and using the equation of state for D and Q proposed by Kamlet-Jacobs[26] and Cooper.[21]

Finally, the standard enthalpy of formation is a key parameter in thermochemistry. The precision of calculations could be improved using a rotating bomb calorimeter for the types of fuels used in non-ideal explosive compositions. Examples of this would be ground coffee (used in the 7/7 bombings in London[74]) or the type of diesel commonly found in fuel stations. However, combustion calorimetry is hugely difficult to conduct since the precision in experimentation needs to be ±0.01–0.02% to have an uncertainty in an enthalpy of formation[75] of ±1 kJ mol^{-1}. Additionally, non-ideal compositions containing sulphur or halogens would require specific testing conditions given the highly erosive nature of the gaseous products. Therefore, it may be prudent to consider those fuels with the lower enthalpy values if the precise detail of the hydrocarbon (for example) cannot be determined.

5 Conclusion

The heat of explosion of non-ideal explosives has been historically difficult to model due to the significant departure of highly non-ideal detonations from idealised one-dimensional wave propagation. A simple method has been introduced to predict the heat of explosion for both $C_aH_bN_cO_d$ explosives, and explosives comprising mixtures or organic, inorganic and organometallic compounds. This method is useful to the first responder since an in-depth knowledge of the chemical or thermochemical properties of explosives is not required. Compared to TNT equivalence criteria published in literature, this procedure combines accuracy and simplicity to predict the performance of a wide range of energetic materials based on the valence of the atoms that exist within any composition, and at what stage reactions take place. The model also allows the thermal energy release sequence to be predicted under ideal conditions should there be excess fuel in the composition and sufficient atmospheric oxygen to contribute to quaternary blast injuries.

Variance from reported literature does exist, but at a scale much reduced from other methods of TNT equivalence (no more than ±15% in examples as opposed to ±30%). This variance is a consequence of the unpredictable chemistry accompanying non-ideal detonations within the detonation and non-detonation reaction zones (for example the role of barium nitrate in Baratol and Baronal), or the enthalpies of formation that are chosen for complex hydrocarbons or polymers such as vaseline, diesel and sawdust. This paper demonstrates that the variance tends towards over-compensation, which from a risk and risk management perspective would be considered as low as reasonably practical.

As such, the valence equation can be used to determine the risk posed by HME as part of predictive performance analysis, therefore allowing evacuation distance for blast and fragmentation to be accurately determined. A general equation for approximate HME performance for a first responder can therefore be developed as a hand-held Application, providing timely levels of assurance whilst more complex interventions follow behind.

Appendix A

A.1 Baseline calculations for FEFO – $C_5H_6N_4O_{10}F_{2(s)}$

$$\Omega = 15.999 \times \left(\frac{d - 2a - \frac{b}{2} + \frac{f}{2}}{\text{molecular weight}} \right) \times 100$$

$$\therefore \Omega = 15.999 \times \left(\frac{10 - 10 - \frac{6}{2} + 1}{320.12} \right) \times 100 = -9.99\%$$

$$\therefore \text{ products of detonation} \rightarrow \frac{c}{2}N_2 + f\text{HF} + \left(\frac{b-f}{2}\right)H_2O$$
$$+ \left(2a - d + \frac{b-f}{2}\right)CO + \left(d - a - \frac{b-f}{2}\right)CO_2$$

$$\therefore C_5H_6N_4O_{10}F_2 \rightarrow \frac{4}{2}N_2 + 2\text{HF} + \left(\frac{6-2}{2}\right)H_2O$$
$$+ \left(10 - 10 + \frac{6-2}{2}\right)CO + \left(10 - 5 - \frac{6-2}{2}\right)CO_2$$

$$\therefore C_5H_6N_4O_{10}F_{2(s)} \rightarrow 2N_{2(g)} + 2\text{HF}_{(g)} + 2H_2O_{(g)} + 2CO_{(g)} + 3CO_{2(g)}$$

A.2 Baseline calculations for PBX 9407 – $C_{1.41}H_{2.66}N_{2.54}O_{2.69}Cl_{0.07}F_{0.09(s)}$

$$\Omega = 15.999 \times \left(\frac{d - 2a - \frac{b}{2} + \frac{f}{2} + \frac{cl}{2}}{\text{molecular weight}} \right) \times 100$$

$$\therefore \Omega = 15.999 \times \left(\frac{2.69 - 2(1.41) - \frac{2.66}{2} + \frac{0.09}{2} + \frac{0.07}{2}}{102.42} \right) \times 100$$

$$\therefore \Omega = 15.999 \times (-0.0135) \times 100 = -21.58\%$$

$$\therefore C_{1.41}H_{2.66}N_{2.54}O_{2.69}Cl_{0.07}F_{0.09}$$

$$\rightarrow \frac{c}{2}N_2 + fHF + clHCl + \left(\frac{b-f-cl}{2}\right)H_2O + \left(2a - d + \frac{b-f-cl}{2}\right)CO$$

$$+ \left(d - a - \frac{b-f-cl}{2}\right)CO_2$$

$$\therefore C_{1.41}H_{2.66}N_{2.54}O_{2.69}Cl_{0.07}F_{0.09}$$

$$\rightarrow \frac{2.54}{2}N_2 + 0.09\,HF + 0.07HCl + \left(\frac{2.66-0.09-0.07}{2}\right)H_2O$$

$$+ \left(2.82 - 2.69 + \frac{2.66-0.09-0.07}{2}\right)CO$$

$$+ \left(2.69 - 1.41 - \frac{2.66-0.09-0.07}{2}\right)CO_2$$

$$\therefore C_{1.41}H_{2.66}N_{2.54}O_{2.69}Cl_{0.07}F_{0.09(s)}$$

$$\rightarrow 1.27N_{2(g)} + 0.09HF_{(g)} + 0.07HCl_{(g)} + 1.25H_2O_{(g)} + 1.38CO_{(g)}$$

$$+ 0.03CO_{2(g)}$$

Appendix B

B.1 Baseline calculations for ammonium nitrate/aluminium (88/12)

The aluminium competes with the detonation products formed, in this case $H_2O_{(g)}$ and any excess oxygen remaining upon formation of the detonation products. As this is a non-ideal explosive, the start point is eqn (19) to determine the oxygen balance for 1 g of composition (Table 11):

$$\therefore \Omega = 15.999 \times \left(\frac{\begin{matrix} d - 2a - \frac{b}{2} - \frac{Li}{2} - \frac{Na}{2} - \frac{K}{2} + \frac{F}{2} + \frac{Cl}{2} + \frac{Br}{2} + \frac{I}{2} \\ -Mg - Ba - Zn - Sr - \frac{3Al}{2} - \frac{3B}{2} - 2Pb \end{matrix}}{1}\right) \times 100 \qquad (19)$$

$$\therefore \Omega = 15.999 \times \left(\frac{d - 2a - \frac{b}{2} - \frac{3Al}{2}}{1}\right) \times 100$$

$$\therefore \Omega = 15.999 \times \left(\frac{0.032982 - \frac{0.043976}{2} - \frac{3}{2}(0.004447)}{1}\right) \times 100$$

$$\therefore \Omega = +6.92\%$$

Table 11 Molecular formula calculation for AN-Al (88/12).[a]

Molecule	Empirical formula	Proportion by mass (g)	Molecular weight (g)	Molar proportions in 1 g	C	H	N	O	Al
Ammonium nitrate	NH_4NO_3	0.88	80.043480	0.010994	0	4	2	3	0
Aluminium	Al	0.12	26.981540	0.004447	0	0	0	0	1
Totals		1		0.015442	0	0.043976	0.021988	0.032982	0.004447

[a] Note: owing to the small molar proportions in 1 g of explosive, up to six decimal places are used in calculations where required.

The heat of explosion for this reaction is then determined from Hess' law (see eqn (24)), with enthalpy values taken from Table 4:

$$\Delta H^{\theta}_{\text{reaction}} = \sum \Delta H^{\theta}_{\text{f(products)}} - \sum \Delta H^{\theta}_{\text{f(reactants)}} \tag{24}$$

$$\therefore \ H_{0.043976}N_{0.021988}O_{0.032982}Al_{0.004447}$$
$$\rightarrow \frac{c}{2}N_2 + \left(\frac{b}{2}\right)H_2O + aCO_2 + \left(\frac{2d-b}{4} - a\right)O_2 + alAl \ \{\text{STAGE 1}\}$$

$$\therefore \ H_{0.043976}N_{0.021988}O_{0.032982}Al_{0.004447}$$
$$\rightarrow \frac{0.021988}{2}N_2 + \left(\frac{0.043976}{2}\right)H_2O + \left(\frac{(2 \times 0.032982) - 0.043976}{4}\right)O_2$$
$$+ 0.004447Al \ \{\text{STAGE 1}\}$$

$$\therefore \ H_{0.043976}N_{0.021988}O_{0.032982}Al_{0.004447}$$

$$\rightarrow 0.010994N_2 + 0.021988H_2O + 0.005497O_2 + 0.004447Al \ \{\text{STAGE 1}\}$$

$$\therefore \ Q_{\text{Stage 1}} = \{(0.021988\Delta H_f(H_2O(g))\} - \{0.010994\Delta H_f(AN(s))\}$$

$$\therefore \ Q_{\text{Stage 1}} = \{(0.021988 \times -241.8)\} - \{(0.010994 \times -365.3)\}$$

$$\therefore \ Q_{\text{Stage 1}} = -1.30 \text{ kJ g}^{-1} = -1300 \text{ kJ kg}^{-1}$$

At the end of Stage 1, the aluminium reacts with the product(s) of detonation as follows (see assumption 4 from Table 4):

$$0.021988H_2O + 0.005497O_2 + 0.004447Al$$
$$\rightarrow 0.002224Al_2O_3 + 0.022027H_2O + 0.002160O_2 \ \{\text{STAGE 2}\}$$

$$\therefore \ Q_{\text{Stage 2}} = \{(0.022027\Delta H_f(H_2O(g)) + (0.002224 \times \Delta H_f(Al_2O_3(s))\}$$
$$- \{(0.021988 \times \Delta H_f(H_2O(g))\}$$

$$\therefore Q_{\text{Stage 2}} = \{(0.022027 \times (-241.8)) + (0.002224 \times (-1669.8))\}$$
$$- \{0.021988 \times (-241.8)\}$$

$$\therefore \ Q_{\text{Stage 2}} = \{-9.04\} - \{-5.32\} = -3720 \text{ kJ kg}^{-1}$$

$$\therefore \ Q_{\text{Stage 1}} + Q_{\text{Stage 2}} = (-1300) + (-3720) = -5020 \text{ kJ kg}^{-1}$$

Therefore, the calculated value of Q_e is 101.8% that of the reported literature value. Q_e in kJ mo1^{-1} has been calculated as follows:

$$1 \text{ mol } NH_4NO_3 = 80.04 \text{ g}$$

$$1 \text{ mol Al} = 26.98 \text{ g}$$

$$\therefore \ 1000 \text{ g mixture contains } 880 \text{ g } NH_4NO_3 = 10.99 \text{ mol}$$

$$\therefore \ 1000 \text{ g mixture contains } 120 \text{ g Al} = 4.45 \text{ mol}$$

$$\therefore \text{mol\% } NH_4NO_3 = \left(\frac{10.99}{10.99 + 4.45}\right) \times 100 = 71.2\% = 56.97 \text{ g}$$

$$\therefore \text{mol\% Al} = \left(\frac{4.45}{10.99 + 4.45}\right) \times 100 = 28.8\% = 7.77 \text{ g}$$

$$\therefore 1 \text{ mol} = 64.77 \text{ g}$$

$$\therefore Q_{\text{Stage 1}} + Q_{\text{Stage 2}} = -5020 \times \left(\frac{64.77}{1000}\right) = -325.15 \text{ kJ mol}^{-1}$$

$$\therefore H_{0.043976}N_{0.021988}O_{0.032982}Al_{0.004447(s)}$$
$$\rightarrow 0.010994N_{2(g)} + 0.022027H_2O_{(g)} + 0.00216O_{2(g)} + 0.002224Al_2O_{3(s)}$$

Appendix C

C.1 Baseline calculations for Baronal (barium nitrate/TNT/Al (50/35/15))

The equation for the above non-ideal explosive can be determined from Table 12 as follows:

$$C_{0.01079}H_{0.0077}N_{0.00845}O_{0.02072}Ba_{0.00191}Al_{0.00556(s)}$$

The oxygen balance for Baronal is calculated using eqn (19):

$$\therefore \Omega = 15.999 \times \left(\frac{d - 2a - \frac{b}{2} - \text{Ba} - \frac{3}{2}\text{Al}}{1}\right) \times 100$$

$$\Omega = 15.999 \times \left(\frac{0.02072 - 2(0.01079) - \frac{1}{2}(0.0077) - 0.00191 - \frac{3}{2}(0.00556)}{1}\right) \times 100$$

$$\therefore \Omega = -23.92$$

Baronal is therefore oxygen deficient. The available oxygen cannot completely produce $CO_{2(g)}$ and $H_2O_{(g)}$. Therefore, the products of detonation can be determined from eqn (11), remembering that Ba and Al shall react with the products of detonation in Stage 2:

$$\therefore \text{products of detonation} \rightarrow \frac{c}{2}N_2 + \left(\frac{b}{2}\right)H_2O + \left(a - \left(d - a - \frac{b}{2}\right)\right)CO + \left(d - a - \frac{b}{2}\right)CO_2$$

$$\text{Therefore, } Q = \sum \Delta H^{\theta}_{\text{f(products)}} - \sum \Delta H^{\theta}_{\text{f(reactants)}} \qquad (21)$$

Table 12 Non-ideal formula for Baronal (90/10).

		Prop. by mass[a] (g)	M_{wt} (g)	Molar prop. in 1 g[b]	C	H	N	O	Ba	Al
TNT	$C_7H_5N_3O_6$	0.35	227.13	0.001541	7	5	3	6	0	0
Al	Al	0.15	26.98	0.005559	0	0	0	0	0	1
Barium nitrate	$Ba(NO_3)_2$	0.5	261.34	0.001913	0	0	2	6	1	0
Totals		1		0.009014	0.01079	0.0077	0.00845	0.02072	0.00191	0.00556

[a] Proportion by mass (g).
[b] Molar proportions in 1 g.

$$\therefore C_{0.01079}H_{0.0077}N_{0.00845}O_{0.02072}Ba_{0.00191}Al_{0.00556}$$
$$\rightarrow \frac{0.00845}{2}N_2 + \frac{0.0077}{2}H_2O$$
$$+\left(0.01079 - \left(0.02072 - 0.01079 - \frac{0.0077}{2}\right)\right)CO$$
$$+\left(0.02072 - 0.01079 - \frac{0.0077}{2}\right)CO_2$$

$$\therefore C_{0.01079}H_{0.0077}N_{0.00845}O_{0.02072}Ba_{0.00191}Al_{0.00556}$$
$$\rightarrow 0.004225N_2 + 0.00385H_2O + 0.00471CO + 0.00608CO_2$$
$$+ 0.00191Ba + 0.00556Al \ \{STAGE\ 1\}$$

$$\therefore Q_{\text{Stage 1}} = \{(0.00385\Delta H_f(H_2O)) + (0.00471\Delta H_f(CO))$$
$$+ (0.00608\Delta H_f(CO_2))\} - \{(0.001541 \times \Delta Hf(TNT))$$
$$+ (0.001913\Delta H_f(Ba(NO_3)_2) + (0.005559\Delta H_f(Al))\}$$

$$\therefore Q_{\text{Stage 1}} = \{(0.00385 \times (-241.8)) + (0.00471 \times (-110.5))$$
$$+ (0.00608 \times (-393.5))\} - \{(0.001541 \times (-62.07))$$
$$+ (0.001913 \times (-998)) + (0.005559 \times 0)\}$$

$$\therefore Q_{\text{Stage 1}} = \{-0.931 - 0.520 - 2.392\} - \{-0.0956 - 1.91\} = -1.837 \text{ kJ g}^{-1}$$

$$\therefore Q_{\text{Stage 1}} = -1837 \text{ kJ kg}^{-1}$$

$$\therefore 0.0042N_2 + 0.00385H_2O + 0.00471CO$$
$$+ 0.00608CO_2 + 0.00191Ba + 0.00556Al$$
$$\rightarrow 0.0042N_2 + 0.00385H_2O + 0.00471CO + 0.000955CO_2$$
$$+ 0.005125C + 0.00191BaO + 0.00278Al_2O_3 \ \{STAGE\ 2\}$$

$$\therefore Q_{\text{Stage 2}} = \{(0.00385\Delta H_f(H_2O) + (0.00471\Delta H_f(CO)) + (0.000955\Delta H_f(CO_2))$$
$$+ (0.00191\Delta H_f(BaO)) + (0.00278\Delta H_f(Al_2O_3))\}$$
$$- \{(0.00385\Delta H_f(H_2O)) + (0.00471\Delta H_f(CO))$$
$$+ (0.00608\Delta H_f(CO_2))\}$$

However, the ΔH_f for H_2O and CO cancel out, as their quantities have not been altered:

$$\therefore Q_{\text{Stage 2}} = \{(0.000955\Delta H_f(CO_2)) + (0.00191\Delta H_f(BaO))$$
$$+ (0.00278\Delta H_f(Al_2O_3))\} - \{(0.00608\Delta H_f(CO_2))\}$$

$$\therefore Q_{\text{Stage 2}} = \{(0.000955 \times (-393.5)) + (0.00191 \times (-548.1))$$
$$+ (0.00278 \times (-1669.8))\} - \{(0.00608 \times (-393.5))\}$$

$$\therefore Q_{\text{Stage 2}} = \{(-0.376) + (-1.047) + (-4.642)\} - \{(-2.39)\}$$

$$\therefore Q_{\text{Stage 2}} = \{(-6.065) - (-2.39)\} = -3.675$$

$$\therefore Q_{\text{Stage 2}} = -3675 \text{ kJ kg}^{-1}$$

$$\therefore Q_{\text{Stage 1}} + Q_{\text{Stage 2}} = (-1837) + (-3675) = -5512 \text{ kJ kg}^{-1}$$

$$\therefore \text{TNT}_e = \frac{-5512}{-4091} = 1.34$$

Therefore, the calculated value of Q_e is 114% that of the reported literature value. Q in kJ mol^{-1} can be calculated as follows:

$$1 \text{ mol Ba(NO}_3)_2 = 261.337 \text{ g}$$

$$1 \text{ mol TNT} = 227.13 \text{ g}$$

$$1 \text{ mol Al} = 26.98 \text{ g}$$

$$\therefore 1000 \text{ g mixture contains } 500 \text{ g Ba(NO}_3)_2 = 1.91 \text{ mol}$$

$$\therefore 1000 \text{ g mixture contains } 350 \text{ g TNT} = 1.54 \text{ mol}$$

$$\therefore 1000 \text{ g mixture contains } 150 \text{ g Al} = 5.55 \text{ mol}$$

$$\therefore \text{mol\% Ba(NO}_3)_2 = \left(\frac{1.91}{1.91 + 1.54 + 5.55}\right) \times 100 = 21.2\% = 55.4 \text{ g}$$

$$\therefore \text{mol\% TNT} = \left(\frac{1.54}{1.91 + 1.54 + 5.55}\right) \times 100 = 17.1\% = 38.84 \text{ g}$$

$$\therefore \text{mol\% Al} = \left(\frac{5.55}{1.91 + 1.54 + 5.55}\right) \times 100 = 61.7\% = 16.64 \text{ g}$$

$$\therefore 1 \text{ mol} = 110.88 \text{ g}$$

$$\therefore Q_{\text{Stage 1}} + Q_{\text{Stage 2}} = -5512 \times \left(\frac{110.88}{1000}\right) = -611.17 \text{ kJ mol}^{-1}$$

$$\therefore \text{C}_{0.01079}\text{H}_{0.0077}\text{N}_{0.00845}\text{O}_{0.02072}\text{Ba}_{0.00191}\text{Al}_{0.00556(s)}$$

$$\rightarrow 0.004225\text{N}_{2(g)} + 0.005125\text{C}_{(s)} + 0.00385\text{H}_2\text{O}_{(g)} + 0.00471\text{CO}_{(g)}$$

$$+ 0.000955\text{CO}_{2(g)} + 0.00191\text{BaO}_{(s)} + 0.00278\text{Al}_2\text{O}_{3(s)}$$

References

1 P. Bergen and D. Sterman, *Jihadist Terrorism 17 years after 9/11 – A threat assessment*, New America, New York, 2018.

2 D. Barten, H. De Cauwer, L. Mortelmans and G. Ciottone, *Prehospital Disaster Med.*, 2022, **37**(1), 19.

3 Chemical and Explosives Terrorism, https://www.interpol.int/en/Crimes/Terrorism/Chemical-and-Explosives-terrorism (accessed July 2024).

4 G. P. Collett, M. Ladyman, R. Hazael and T. Temple, The use of a predictive threat analysis to propose revisions to existing risk assessments for precursor

chemicals used in the manufacture of home-made explosives (HME), *Heliyon*, 2021, 7, DOI: 10.1016/j.heliyon.2021.e08343.
5 B. Cochrane, *The development of the British approach to improvised explosive device disposal in Northern Ireland*, Cranfield University, Shrivenham, 2015.
6 A. Pettitt, *The historical use, development and influence of improvised explosive devices*, PhD thesis, Cranfield University, 2023.
7 J. DooJin, K. KiTae and H. SangEul, *Computation*, 2017, **5**(3) DOI: 10.3390/computation5030041.
8 International Ammunition Technical Guidelines, https://unsaferguard.org/un-saferguard/kingery-bulmash (accessed February 2023).
9 R. Meyer, J. Kohler and A. Homburg, *Explosives – Sixth, Completely Revised Edition*, Wiley VCH, 2007.
10 B. T. Fedoroff and O. E. Sheffield, *Encyclopedia of Explosives and Related Items*, U.S Army Research and Development Coommand, Picatinny Arsenal, New Jersey, 1962, vol. 2.
11 N. Phokion, *Nitroglycerine and Nitroglycerine Explosives*, The Williams & Wilkins Company, Hamburg, 1928.
12 J. Ledgard, *The Preparatory Manual of Explosive*, Uvkchem, 4th edn, 2014.
13 J. C. Oxley, J. L. Smith and L. McLennan, Homemade Explosives, in *Counter Terrorist Detection Techniques of Explosives*, Elsevier B.V., Amsterdam, 2022, ch. 11, vol. 2, pp. 383–422.
14 GICHD, *The IED Clearance Good Practice Guide*, GICHD, Geneva, 2021, ch. 5, p. 1.
15 Deregulation Act 2015 – Schedule 21, https://www.legislation.gov.uk/ukpga/2015/20/schedule/21/enacted (accessed February 2021).
16 R. V. Eastin and G. L. Arbogast, *Demand and Supply Analysis: Introduction*, CFA Institute, Los Angeles, 2011, ch. 13.
17 Afghan bomb makers shifting to new explosives for IEDs, https://eu.usatoday.com/story/news/world/2013/06/25/ammonium-nitrate-potassium-chlorate-ieds-afghanistan/2442191/ (accessed 19 July 2024).
18 Y. B. Zeldovich, *Zh. Exp. Teor. Fiz.*, 1940, **10**, 542–568.
19 J. von Neumann, *Progress Report on the Theory of Detonation Waves*, OSRD, Princeton, 1942.
20 W. Doering, *Ann. Phys.*, 1943, **43**, 421–436.
21 P. W. Cooper, *Explosives Engineering*, VCH Publishers Inc., New York, 1996.
22 C. L. Mader, *Numerical Modelling of Explosives and Propellants*, CRC Press, Boca Raton, 2013.
23 J. H. Lee, *The Detonation Phenomenon*, Cambridge University Press Cambridge, 2008.
24 J. Paszula, W. A. Trzcinski and K. Sprzatczak, *Cent. Eur. J. Energ. Mater.*, 2008, **5**(1), 3–11.
25 S. P. Sivapirakasam, N. V. Kumar, G. Jeyabalaganesh and K. Nagarjuna, *Combust., Explos. Shock Waves*, 2022, **57**, 726–735.
26 M. J. Kamlet and S. J. Jacobs, *J. Chem. Phys.*, 1968, **48**, 23–55.
27 J. Akhavan, *The Chemistry of Explosives*, RSC Publishing, Cambridge, 3rd edn, 2011.
28 M. H. Keshavarz and H. R. Pouretedal, *Thermochim. Acta*, 2004, **414**(2), 203–208.
29 D. Scott-Stewart and A. R. Kasimov, *J. Appl. Math.*, 2005, **66**(2), 384–407, https://www.jstor.org/stable/4096119.
30 J. Radun and Z. Bajic, *Sci.-Tech. Rev.*, 2006, **56**(1), 58–61.
31 S. Elbasuney, M. Gaber Zaky, M. Radwan, M. Bennaya and S. M. Abdelkhalek, *IOP Conf. Ser.: Mater. Sci. Eng.*, 2020, **975**, 012008.
32 B. Khasainov, B. Ermolaev and H. Presles, presented in 12th Symposium (International) on Detonation, Numerical modelling of non-ideal

detonations in ammonium nitrate/aluminium mixtures and their blast effect, San Diego, 2002.
33 X. Mao, Y. Zheng, F. Shen, Y. Wang, X. Wang and Y. Li, *J. Energ. Mater.*, 2022, **40**(2), 182–205.
34 P. P. Vadhe, *Combust., Explos. Shock Waves*, 2008, **44**(4), 461–477.
35 X. Fan, L. Zhang and X. Wang, *J. Phys.: Conf. Ser.*, 2021, **1721**, 1–13.
36 P. Brouseau and C. J. Anderson, *Propellants, Explos., Pyrotech.*, 2002, **27**(5), 300–306.
37 W. A. Trzcinski, S. Cudzilo and J. Paszula, *Propellants, Explos., Pyrotech.*, 2007, **32**(6), 502–508.
38 D. Sherwood and P. Dalby, *Enthalpy and thermochemistry, Modern Thermodynamics for Chemists and Biochemists*, Oxford Academic, Oxford, 2018.
39 C. M. Tarver, Detonation Reaction Zones in Condensed Explosives, in *14th APS Topical Conference on SCCM*, Livermore, 2005.
40 C. E. Anderson, W. E. Baker, D. K. Wauters and B. L. Morris, *Int. J. Mech. Sci.*, 1983, **25**(6), 455–464.
41 Y. Sun and J. Hui, *The Military Mixed Explosive*, Industry Press, Beijing, 1995.
42 A. Hahma, K. Palovuori and Y. Solomon, TNT-equivalency of thermobaric explosive, in Energetic Materials, *Performance & Safety,* 36th International Annual Conference of ICT & 32th International Pyrotechnics Seminar, Karlsruhe, Germany, 2005.
43 Y. Junzheng, *Energy release law and assessment on blast wave power of aluminized explosives*, Beijing Institute of Technology, Beijing, 2017.
44 S. K. Kolev and T. T. Tsonev, *Propellants, Explos., Pyrotech.*, 2021, **47**(2), e202100195.
45 A. Beavan and P. Parker, *Surgery*, 2018, **36**(8), 394–400.
46 S. Loebeckke and G. Collett, *Efficacy of FICT explosive dtatabases determining HME products of detonation*, Cranfield University, Swindon, Wiltshire, 2022.
47 J. C. Oxley, J. L. Smith, M. Donnelly and M. Porter, *Int. J. Energ. Mater. Chem. Propul.*, 2014, **13**(6), 517–557.
48 C. N. Kingery and G. Bulmash, *Air Blast Parameters from TNT Spherical Air Burst and Hemishperical Surface Burst*, Ballistic Research Laboratories, Aberdeen, 1984.
49 R. H. Cole, *Underwater Explosions*, Dover Publications, New York, 1948.
50 P. M. Locking, The Trouble with TNT Equivalence, in *26th International Symposium on Ballistics*, Miami, Florida, 2011.
51 R. K. Wharton, S. A. Formby and R. Merrifield, *J. Hazard. Mater.*, 2000, **79**(1–2), 31–39.
52 L. V. De Yong and G. Campanella, *J. Hazard. Mater.*, 1989, **21**(2), 125–133.
53 P. Locking, TNT Equivalence, Berthelot Theory and the Trauzl Lead Block Test, in *The 32nd International Symposium on Ballistics*, Reno, Nevada, 2022. https://dpi-proceedings.com/index.php/ballistics32/article/view/36049.
54 Headquarters, U.S. Army Materiel Command, *Engineering Design Handbook – Properties of Explosives of Military Interest*, Vols. AMCP 706-177, U.S. army Materiel Command, Washington, 1971.
55 D. Dusenberry, *Handbook for Blast Resistant Design of Buildings*, Wiley, New York, 2010.
56 G. Bikelyte, M. Hartel, J. Stierstorfer, T. M. Klapotke, A. A. Pimerzin and S. P. Verevkin, *J. Chem. Thermodyn.*, 2017, **111**, 271–278.
57 B. M. Dobratz, *Properties of Chemical Explosives and Explosive Simulants*, Vols. UCRL-51319/1, Lawrence Livermore Laboratory, Livermore, 1972.
58 J. B. Pedley, R. D. Naylor and S. P. Kirby, *Thermochemical Data of Organic Compounds*, Chapman and Hall, New York, 2nd edn, 1986.

59 Heats of Formation of Compounds, https://www.thoughtco.com/common-compound-heat-of-formation-table-609253 (accessed) 30 May 2023.
60 J. Li and A. S. Morgans, *Fuel*, 2016, **184**, 735–748.
61 A. Marshall, *Explosives*, P. Blakiston's Son & Co, Philadelphia, 2nd edn, 1917.
62 J. B. Pedley, R. N. Naylor and S. P. Kirby, *Thermochemical Data of Organic Compounds*, Chapman and Hall, London, 2nd edn, 1986.
63 R. Matyas and J. Pachman, *Primary Explosives*, Springer, Berlin, 2013.
64 NASEM, *Reducing the threat of IED attacks by restricting access to explosive precursor chemicals*, The National Academies Press, Washington DC, 2018.
65 W. E. Garner and C. L. Abernethy, *Heat of Combustion and Formation of Nitro-Compunds. Part 1. Benzene, Toluene, Phenol and Methylaniline Series.*, Royal Society, London, 1920.
66 P. W. Cooper, Comments on TNT Equivalence, in *20th International Pyrotechnics Seminar*, Colarado Springs, 1994.
67 A. Biessikirski, D. Czerwonka, J. Bieganska, L. Kuterasinski, M. Ziabka, M. Dworzak and M. Twardosz, *Energies*, 2021, **14**(172), 1–15.
68 A. Maranda, A. Paplinski and D. Galezowski, *J. Energ. Mater.*, 2003, **21**(1), 1–13.
69 B. T. Fedoroff and O. E. Sheffield, *Encyclopedia of Explosives and Related Items*, U.S Army Research and Development Coommand, Picatinny Arsenal, New Jersey, 1962, vol. 2.
70 M. W. Smith, D. Shekhawat and D. A. Berry, *Fuel Cells: Technologies for Fuel Processing*, Elsevier, 2011.
71 J. W. Kury and R. D. Breithaupt, Detonation Products Equation of State for Baratol, in *9th Symposium (International) on Detonation*, Portland, Oregon, 1989.
72 R. F. Chaiken and J. F. Edwards, *Detonation and Critical Diameter of Heterogeneous Explosives*, Pergamon Press, Pittsburgh, 1976, vol. 3.
73 D. Price, Contrasting patterns in the behavior of high explosives, in *Eleventh Symposium (International) on Combustion*, Pittsburgh, 1967.
74 Report of the 7 July Review Committee, Greater London Authority, https://www.london.gov.uk/sites/default/files/gla_migrate_files_destination/archives/assembly-reports-7july-report.pdf (accessed 12 June 2023).
75 M. V. Roux and M. Temprado, in *Handbook of Thermal Analysis and Calorimetry*, Elsevier B.V., Madrid, 2008, ch. 14, vol. 5.

Degradation kinetics of sea-dumped decoy flares

Cpt Francesco Saverio Amendola,* Lt Col Silvana Grossi and Lt Simona Pergolini

Flares have been the most widely used countermeasure for electronic warfare systems in infrared (IR) band. Some cases of mislight, with consequent dumping, have been reported during flight training sessions over the sea. In order to estimate the degradation time of spectral flares in marine environments a kinetic model has been elaborated, starting from the relative stability of the nitrate esters. The stabiliser depletion, as a function of time and temperature, has been evaluated by a combination of the zero-order and the first order models, for the purpose of achieving a better description of the heterogeneous reactions. Experimental conditions and theoretical considerations suggested that hydrolysis is the relevant mechanism at seawater temperatures and, consequently, the surface area is a deciding factor for the degradation kinetic model. The elaborated equations show that the chemical processes for a complete degradation of flares accidentally released in the Mediterranean Sea may take a time period of several years.

1 Introduction

Traditionally, countermeasure flares, employed to elude even the most advanced missile seeking technologies, are manufactured from pyrotechnic mixtures, able to emulate the emissions of various military platforms.[1,2]

Nowadays, a relevant multiplicity of emergency countermeasure formulations is commercially available and this number is constantly increasing due to the need of more and more sophisticated defence systems. Openly published descriptions on in-service flare compositions are partly limited by commercial and national restrictions.[3] Nevertheless, the scientific literature provides a database of interesting pyrotechnic formulations for countermeasures. A high number of papers describe the particular application of spectral mixtures[4] and, more specifically, of nitrate esters-based mixtures: widespread for the production of smokeless powders, these formulations can also be used in the manufacture of countermeasures, where the physicochemical and explosive properties of nitrocellulose and nitroglycerin (Fig. 1) are usually combined in a double base preparation.[3,5]

Since the nitrate esters-based powders show a relatively low chemical stability, further needed is a stabiliser such as 2-nitrodiphenylamine or ethyl centralite.

The molecular structures of all these additives are characterised by the presence of activated phenyl rings (Fig. 2), acting as scavengers for the nitrate esters decomposition products.[6,7]

Aerospace Test Division, Italian Air Force, Pomezia (RM), Italy.
E-mail: francesco.amendola@aeronautica.difesa.it

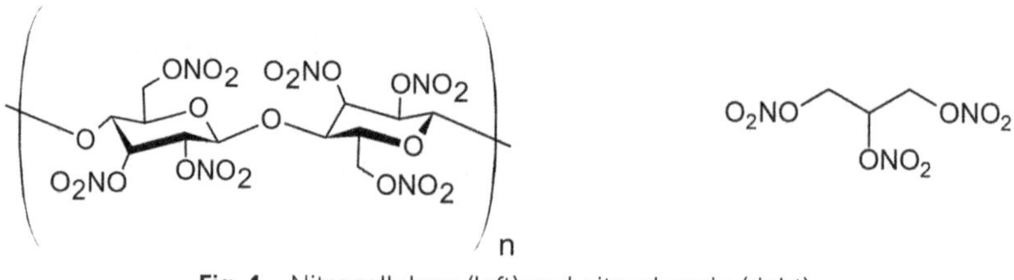

Fig. 1 Nitrocellulose (left) and nitroglycerin (right).

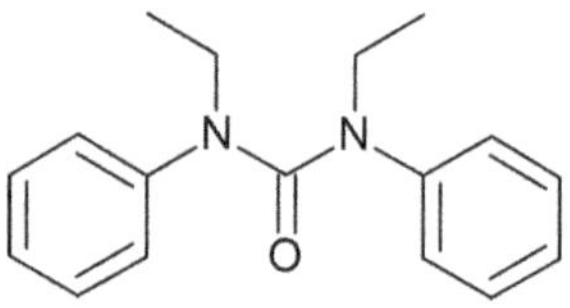

Fig. 2 Ethyl centralite.

1.1 Degradation of nitrate ester-based countermeasures

In principle, there are four main processes contributing to the ageing of nitrate esters-based powders:

- Intrinsic decomposition
- Nitrocellulose chain scission
- Insoluble gas generation
- Nitroglycerin migration

Degradation of nitrate esters starts with the homolytic breaking of the weak nitrate ester bond $O–NO_2$. This scission, so-called *thermolysis* because it is enhanced by high temperatures, leads to the formation of two radical species:[5]

$$RO–NO_2 \rightarrow RO^{\bullet} + {}^{\bullet}NO_2$$

The radical ${}^{\bullet}NO_2$ rapidly reacts with the immediate surroundings, producing further radicals and promoting an autocatalytic decomposition.

$$RO–NO_2 + {}^{\bullet}NO_2 \rightarrow N_2, N_2O, {}^{\bullet}NO_2, {}^{\bullet}NO, H_2O, H_2, CO_2, CO, C_2H_2O_4 \; etc.$$

The presence of water and oxygen also self-accelerates the degradation process: by generating residual acids, H_2O and O_2 pave the way to the neutral to acid hydrolysis of the nitrate esters.[8]

$$RO–NO_2 + 2H_2O \rightarrow ROH + HNO_3$$

$$3\,{}^{\bullet}NO_2 + H_2O \rightarrow 2HNO_3 + {}^{\bullet}NO$$

$$2\,{}^{\bullet}NO + O_2 \rightarrow 2\,{}^{\bullet}NO_2 \rightarrow N_2O_4$$

$${}^{\bullet}NO_2 + {}^{\bullet}NO + H_2O \rightarrow 2HNO_2$$

The enhanced hydrolysis dominates the decomposition at low temperatures. As reported for the thermolysis,[7] the hydrolysis is also followed by heat generation, with consequent risk of autoignition.

In turn, the alkoxy radical $RO^•$ of the nitrocellulose chain performs intra- or intermolecular reactions, producing gaseous, stable molecules.[9]

$$RO^• \rightarrow N_2,\ N_2O,\ {}^•NO_2,\ {}^•NO,\ H_2O,\ H_2,\ CO_2,\ CO,\ C_2H_2O_4\ \textit{etc.}$$
$$RO{-}NO_2 + RO^• \rightarrow N_2,\ N_2O,\ {}^•NO_2,\ {}^•NO,\ H_2O,\ H_2,\ CO_2,\ CO,\ C_2H_2O_4\ \textit{etc.}$$

The gaseous products can accumulate within the flare pellet and lead to physical absorption or gas cracking.

Stabilisers reduce the risk of gas accumulation and, more generally, slow down the autocatalytic decomposition: notwithstanding the stabilisers cannot suppress the homolytic breaking of the weak bond $O{-}NO_2$, they can slow down the subsequent reactions, scavenging the radicals ${}^•NO_x$ and the acidic species. This property allows to reduce the autocatalysis, limiting as a consequence the generation of gases and heat.[10]

$$S + {}^•NO_x \rightarrow S{-}NO_x$$

$$S + HNO_3 \rightarrow S{-}NO_2 + H_2O$$

$$S + HNO_2 \rightarrow S{-}NO + H_2O$$

The succession of the last three reactions involves the stabiliser depletion. Once the stabilisation factor is exhausted, the nitrate esters will undergo a fast decay.

The principal decomposition phenomena are summarised in Fig. 3. With the exception of the nitroglycerin migration, all the reactions occur between substances in different phases and can be ascribed in the class of the *heterogeneous processes.*

1.2 Degradation kinetics principles

A realistic estimation of the time needed for the deactivation of decoy flares in marine environments is the main objective of this work. The deactivation can be considered complete when the dumped countermeasure becomes unable to damage people and the immediate surroundings, even if rescued and intentionally or accidentally exposed to ignition stimuli.

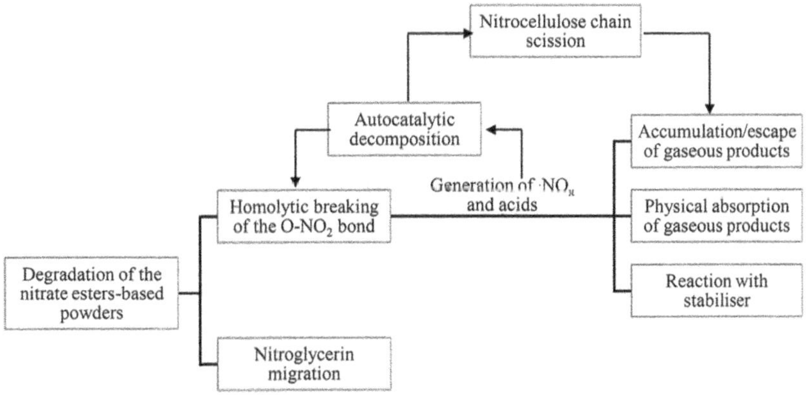

Fig. 3 Degradation of the nitrate esters-based mixtures.

This concept can be considered a unique in the scientific literature: whilst a large number of papers related to the fate of sea-dumped munitions and their toxic effects has been published,[11–13] the risks that come from the activation of rescued explosive articles and, in particular, the time needed for these dangerous properties to reduce have always been a marginal topic, and this is even more true if the discussion focuses on propellants and pyrotechnics.

The rationale behind this work includes the hypothesis that the countermeasures are dumped in coastal waters, where temperature, oxygen concentration and probability of rescue are higher.

In order to define the degradation kinetics of the sea-dumped spectral flares, the stabiliser depletion has been considered an essential factor: following the depletion rate is an indirect, but simple and efficient method for monitoring the nitrate esters decomposition. The turning point is reached when the stabiliser concentration equals zero: until the stabiliser is present, the energetic ingredients are preserved and the countermeasure can be considered still active and dangerous; when the stabiliser will be completely depleted, the nitrate esters degradation will become so fast that the flare pellet can deactivate very quickly.[14–16]

The stabiliser depletion rate – in this specific case ethyl centralite (EC) – has been evaluated starting from the simple linear and exponential decreases[17,18] and combining them in a complex model, suitable to describe heterogeneous processes.[15,19,20] The linear and the exponential decreases are respectively rationalised by the zero-order (eqn (1)) and the first-order kinetics (eqn (2)):

$$\frac{\mathrm{d}S(t,T)}{\mathrm{d}t} = -k_0(T) \tag{1}$$

$$\frac{\mathrm{d}S(t,T)}{\mathrm{d}t} = -k_1 S(t,T) \tag{2}$$

The ethyl centralite percentage $S(t,T)$ is a time- and temperature-dependent factor. The zero-order and the first-order rate constants k_0 and k_1 are related to the activation energy and the temperature of the processes by the Arrhenius equation (eqn (3) and (4)):

$$k_0 = A_0 \exp\left(-\frac{Ea_0}{RT}\right) \tag{3}$$

$$k_1 = A_1 \exp\left(-\frac{Ea_1}{RT}\right) \tag{4}$$

The kinetics of heterogeneous reactions can be better describe combining the linear and the exponential approaches,[21,22] resulting in

$$\frac{\mathrm{d}S(t,T)}{\mathrm{d}t} = -k_1 S(t,T) - k_0(T) \tag{5}$$

Integrating the eqn (5) with respect to time, it is possible to evaluate the remaining EC percentage after the decoy flare has been immersed in seawater at temperature T for a time t:

$$S(t,T) = \left(S(t_0) + \frac{k_0(T)}{k_1(T)}\right)\exp[-k_1(T)\cdot t] - \frac{k_0(T)}{k_1(T)} \tag{6}$$

where $S(t_0)$ is the EC percentage at the initial time t_0.

Rearranging the eqn (6), the time necessary to reach a specific EC decrease can be calculated:

$$t_{S(t,T)} = \frac{1}{k_1(T)}\ln\left(\frac{1+\dfrac{k_0(T)}{S(t_0)\cdot k_1(T)}}{\dfrac{S(t,T)}{S(t_0)}+\dfrac{k_0(T)}{S(t_0)\cdot k_1(T)}}\right) \tag{7}$$

For the complete stabiliser depletion, the eqn (7) can be reduced in the following form:

$$t_{S(t,T)=0} = \frac{1}{k_1(T)}\ln\left(\frac{S(t_0)\cdot k_1(T)}{k_0(T)}+1\right) \tag{8}$$

In order to evaluate the adequacy of the proposal, the kinetic results can be compared with the ageing estimation achievable with other predictive kinetic method, for example by applying the NATO AOP-48 procedure.[23]

To clarify, all the kinetics described in this paragraph are summarised in Table 1.

1.3 Experimental determination of the kinetic parameters

The reaction order cited in the previous paragraph is a purely experimental concept: its value – that can be positive, negative or zero – allows to describe a kinetics by simple equations, such as those shown in Table 1. Experimentally, the stabiliser percentage is measured for samples both at the initial time and after ageing tests in seawater. Tests need to be conducted at different temperatures and increasing ageing times: reporting the stabiliser depletion as a function of time allows to obtain the k_0 and k_1 values for any test temperature.

Table 1 Reaction kinetics.

Chemical kinetics	Rate equation	Description
Zero order	$\frac{dS(t)}{dt} = -k_0$	Linear decrease
First order	$\frac{dS(t)}{dt} = -k_1 \cdot S(t)$	Exponential decrease
Combination of zero and first order	$\frac{dS(t)}{dt} = -[k_1 \cdot S(t) + k_0]$	Combination of linear and exponential order
nth order	$\frac{dS(t)}{dt} = -k_n \cdot S^n(t)$	Model from NATO AOP-48

The Arrhenius equation will provide the temperature dependence of the rate constants:

The Arrhenius equation provides the temperature dependence of the rate constants:

$$\ln k = \ln A - \frac{E_a}{RT} \tag{9}$$

The slope of the linear interpolation obtained by putting $\ln k$ as a function of $1/T$ is proportional to the activation energy E_a, whereas $\ln A$ is the frequency factor. For heterogeneous processes the activation energy includes not only the component required to form the activated complex, but also a fraction connected to the diffusion of liquids and gases inside the solid matrix of the countermeasure.

Once E_a and A are quantified for both zero- and first-order kinetics, k_0 and k_1 values can be calculated in order to predict, by the eqn (7) and (8), the stabiliser depletion rate at any temperature.

The reliability of the elaborated kinetic model needs to be confirmed by the characterisation of a sea-dumped flare, naturally aged for 334 days. The chemical stability[24] and the heat sensitiveness of naturally aged samples can be assessed and compared with the values detected for a freshly unpacked countermeasure.

The present work focuses on accidental dumping in coastal waters, where temperature, oxygen concentration and probability of rescue are higher. Before carefully describing the experimental activity, it is worth underlining that the kinetic tests and the predictive model are essentially focused on the degradation that is observed when the nitric esters are immersed in a solution at a certain level of salinity. Any other interaction or event that may possibly involve the material under investigation and that is not related to the chemical degradation (*e.g.* mechanical or biological phenomena) is not taken into consideration.

2 Experimental section

2.1 Materials

The degradation of flares in a marine environment was simulated in a solution of sodium chloride (ACS reagent, ≥99.0%) and sodium hydroxide (ACS reagent, ≥97.0%). Samples for the HPLC analysis were prepared using acetonitrile for UHPLC analysis (Supergradient, ACS) and a freshly prepared 2% calcium chloride (≥90.0%) solution; the same acetonitrile was used, in solution with double-distilled water, for the chromatographic runs. The quantification of ethylcentralite was carried out diluting a standard solution, freshly prepared from the 99% pure reagent.

2.2 Methods

2.2.1 Accelerated ageing. Sessions of accelerated ageing were performed using an IKA Labortechnik RETbasic heating magnetic stirrer equipped with an IKA Werke ETS-D4 fuzzy digital thermometer. Samples with a length of 1 cm and a mass between 0.7 and 0.85 g were immersed in a solution having salinity and pH comparable with the marine parameters.[25]

The saline solution was prepared by dissolving, in a 1000 ml flask, 35 g of NaCl in one liter of double-distilled water. Before the solution is finally brought to volume, the pH is corrected by adding a few drops of a dilute NaOH solution.

Tests were conducted at temperatures of 66.5, 71.5, 76.5 and 84.5 °C. The samples were placed in a 500 ml single neck round-bottom flask containing a volume of 300 ml of saline solution and a magnetic bar for stirring; a bubble condenser is placed on the neck of the flask. The reaction flask is heated in a silicone oil bath and conditioned at the test temperature for two hours. Appropriate agitation of the solution is guaranteed by a rotational speed of 500 rpm. For each set temperature, the samples were immersed for 24, 48, 72 and 96 hours. At the end of the established time, the samples were brought back to room temperature and the excess water present on the surface was removed by desiccation.

2.2.2 **Natural ageing.** The natural ageing process was emulated by immersing two countermeasure pellets in a 30 l plastic keg filled with seawater. The keg was then stored, without the lid, in a shelter at room temperature; the volume of seawater in the keg was constantly monitored and topped up when necessary.

2.2.3 **High performance liquid chromatography.** HPLC quantification was performed on both artificially aged samples and freshly unpacked countermeasures. The extraction of ethyl centralite was carried out according to the NATO AOP-48 procedure.[23] Samples were dissolved in acetonitrile using an ASAL mod. 717 mechanical stirrer. Each propellant sample was placed in a 250 ml conical flask containing a volume of acetonitrile proportional to the initial mass of the sample and mechanically stirred until completely dissolved. The subsequent addition of 2% calcium chloride $CaCl_2$ solution promoted the precipitation of nitrocellulose. The quantities of solvent referred to the proportions in the NATO AOP-48 procedure, where 250 ml of acetonitrile and 50 ml of 2% $CaCl_2$ solution are suggested for each gram of launch powder. The solution is left to stand for an hour. Before the injection, 2 ml of supernatant is filtered through a nylon syringe filter with a diameter of 13 mm and a porosity of 0.22 μm.

The standard solution was prepared according to the following process: 50.8 mg of ethyl centralite was weighed and transferred to a 50 ml volumetric flask and diluted to the appropriate volume with acetonitrile. This solution included 1016 $mg\,l^{-1}$ of ethyl centralite. Calibration series were prepared in volumetric flasks by the appropriate dilution of stock solution with a mixture of acetonitrile and water in the proportion of 60 : 40 (v : v). The calibration curve was plotted with five concentrations in the range of 50.80–91.44 $\mu g\,mL^{-1}$ of ethyl centralite (Table 2). The experiments were performed in three replicates for each level. The linearity of the calibration curve was evaluated by the linear regression statistics of concentrations against peak area.

Table 2 Calibration series.

	Concentration ($mg\,l^{-1}$)
Standard 1	50.800
Standard 2	60.960
Standard 3	71.120
Standard 4	81.280
Standard 5	91.440
Calibration curve:	$A = 1.94 \times 10^4 \cdot c + 1.27 \times 10^5$ $R^2 = 0.9995$

Table 3 HPLC setup.

Instrumental setup	
Chromatographic column	C18 Phenomenex Luna (with pre-column)
Chromatographic column temperature	35 °C
Elution	Isocratic
Eluent	$H_2O:CH_3CN = 40:60$
Flow	1.00 $ml\,min^{-1}$
Pressure	1290–1350 psi
Injection volume	10 µl
Detection	254 nm

Table 4 Heat flow microcalorimeter setup.

Instrumental setup	
Artificial ageing temperature	80 °C
Test duration	10.58 days
Equivalent time for ageing at 25 °C	10 years
Heat flow limit (STANAG 4582)	113.6 $\mu W\,g^{-1}$

The analysis was performed on a Waters 600 HPLC/GPC chromatographic system, with a 996 photodiode array detector and a reversed phase C18 column (4.6 × 250 mm id., particle size of 5 µm). The chromatographic conditions are reported in Table 3 below.

2.2.4 Heat flow microcalorimetry. Tests were conducted on a TAM 48 microcalorimeter. A Sartorius BP 210s analytical scale was used to weigh the samples. Three samples from a freshly unpacked decoy flare and three samples from a 365-day naturally aged flare were subjected to artificial ageing, monitoring the normalized heat flux throughout the test. As required by the NATO STANAG 4582, the data obtained before the sample released a heat of 5 $J\,g^{-1}$ are not taken into consideration as it comes from the aerobic decomposition of the propellant exposed to the air inside the vial at the beginning of the ageing.[24]

The test was conducted at 80 °C for 10.58 days, equivalent to an isothermal storage of 10 years at 25 °C. The experimental method is reported in Table 4.

2.2.5 Ignition sensitivity test. The thermal sensitivity of the samples was tested using an IKA Labortechnik RETbasic heating magnetic stirrer equipped with an IKA Werke ETS-D4 fuzzy digital thermometer. The ignition sensitivity test is conducted by subjecting the samples to a temperature ramp of 4 $^{\circ}C\,min^{-1}$.

Three samples from a 365-day naturally aged flare are individually subjected to a controlled heating, in order to evaluate their sensitivity to thermal activation. The same test was conducted on three samples of freshly disassembled flares. All the tested samples had a length of 2 cm.

The controlled heating was carried out by placing each sample in a 32 cm glass test tube. The latter was heated in a silicon oil bath.

3 Results and discussions

3.1 Description of the investigated countermeasure

Tested countermeasures are characterised by a length of 14.7 cm, a width of 4.7 cm and a height of 2.0 cm. The main pellet made up of 4 inner rods and 4 corner rods; they are collocated to form a cavity (Fig. 4), filled with filaments made out of the same material as the main pellet.

In order to facilitate a homogeneous ignition, a primer based on boron and potassium nitrate is dispersed over the entire countermeasure. The total mass of the energetic materials is 133.5 g on average.

3.2 Natural ageing

The most noticeable and rapid effect of the natural ageing was the almost complete dissolution of the primer in a short time. No significant change affected the main pellets, except a slight variation in colour. The nitrate esters-based filaments dispersed and fragmented easily; in most cases, they were difficult to recover from the keg.

3.3 Artificial ageing

The saline solution used to emulate the sea water was characterized by a salinity of $3.5 \pm 0.1\%$ and a pH of 8.0 ± 0.1. In order to verify the variability in degradation of the different rods, the tests at 66.5 and 76.5 °C

Fig. 4 Nitrate esters-based countermeasure.

were conducted on samples from an inner rod, while the tests at 71.5 and 84.5 °C were performed on samples from a corner rod.

At the end of all ageing treatments, no macroscopic variation was observed on the samples, either in size or mass.

3.4 HPLC analysis

For the ethyl centralite quantification, the area of the chromatographic peak was quantified by setting the detector wavelength at 254 nm. The stabiliser had a retention time of 13.043 nm. The quantified percentages are reported in Tables 5 and 6 below:

3.5 Extrapolation of the kinetic law

The degradation rate was obtained by correlating the decrease in stabilizer over time according to the zero-order and first order kinetics.

It has been calculated that introducing the raw stabiliser percentages $\%_{EC}$ in the kinetic equations returns an unrealistic prediction of the reaction rate, with very short degradation times. The unreliability of this first prediction was also confirmed by comparison with the phenomena observed during the natural ageing. In the case of immersion in the marine environment – and particularly at low temperatures, where the hydrolysis reaction prevails over any other degradation mechanism – the effect that water

Table 5 Ethyl centralite percentages from samples of a freshly unpacked flare.

Sample	Rod	$m(t_0)$ (g)	C_{EC} (mg l^{-1})	V (ml)	$\%_{EC}$	$\%_{EC\,average}$
A	Inner	0.7472	75.876	224.329	2.278	2.266
B		0.7937	75.202	237.965	2.255	
C	Corner	0.7396	75.586	221.910	2.268	2.258
D		0.7445	74.551	224.561	2.249	

Table 6 Ethyl centralite percentages from artificially aged samples.

T (°C)	Duration (h)	Sample	Rod	m_i (g)	C_{EC} (mg l^{-1})	V (ml)	$\%_{EC}$
66.5	24	1	Inner	0.7793	74.677	233.757	2.240
	48	2		0.8238	74.296	246.882	2.227
	72	3		0.7725	73.506	231.595	2.204
	96	4		0.8474	72.515	254.393	2.177
71.5	24	1	Corner	0.8137	72.122	243.953	2.162
	48	2		0.7403	68.598	222.038	2.057
	72	3		0.7302	68.143	219.109	2.045
	96	4		0.7293	66.880	218.853	2.007
76.5	24	1	Inner	0.7978	72.943	239.488	2.190
	48	2		0.8099	68.984	242.802	2.068
	72	3		0.7169	67.706	215.284	2.033
	96	4		0.7995	67.652	239.872	2.030
84.5	24	1	Corner	0.7164	68.831	215.156	2.067
	48	2		0.7540	66.985	225.991	2.008
	72	3		0.7541	64.879	225.991	1.944
	96	4		0.7381	63.724	221.527	1.913

has on the stability of the nitric esters becomes of prime importance, as reported in par. 1.1. It follows that the extent of the degradation will be affected by the possibility that water diffuses within the sample and, consequently, by the sample portion actually exposed to the surrounding water.

In order to take this influence into account, the stabilizer content, measured for the freshly unpacked samples and at the end of the ageing tests, was normalized by the surface area A_s of the tested sample:

$$\Sigma = \frac{m_{EC}}{A_s} = \frac{\%_{EC} \cdot \bar{m}(t_0)}{100 \cdot A_s} \tag{10}$$

In the elaborated equation, the masses are expressed in g and the surface areas are reported in cm^2. The percentage of ethyl centralite is multiplied by the average mass of the samples at time t_0, in order to obtain Σ values not dependent on the masses of the individual samples.

The A_s values are calculated for both the inner and corner samples. The details are indicated in Tables 7 and 8:

The calculated Σ values are reported in the Tables 9 and 10 below:

The parameter Σ is then used to obtain the rate constants k_0 and k_1. For zero-order kinetics it is sufficient to graph the Σ values *versus* time and extrapolate, as the opposite of the angular coefficient, the rate constant k_0, measured in $g\,h^{-1}\,cm^{-2}$ (Fig. 5, 7, 9 and 11). The first-order kinetics, in turn, requires to plot the natural logarithm of Σ as a function of time: in this case, the opposite of the angular coefficient is representing the rate constant k_1, measured in h^{-1} (Fig. 6, 8, 10 and 12).

Table 7 Top view, dimensions and calculated surface areas of the inner rod.

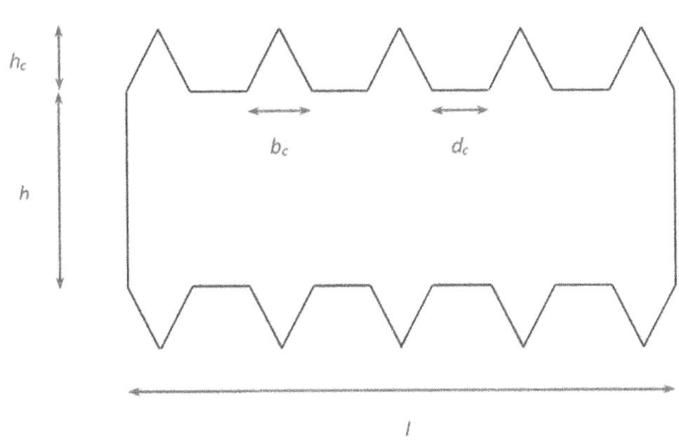

$l = 0.9$ cm
$h = 0.4$ cm
$b_c = 0.1$ cm
$h_c = 0.1$ cm
$d_c = 0.1$ cm

For an inner sample with unit length (L = 1 cm)	$A_s = 2 \cdot (h \cdot l) + 2 \cdot (h \cdot L) + 8 \cdot (d_c \cdot L) + 20 \cdot \left(L \cdot \sqrt{\left(\frac{b_c}{2}\right)^2 + h_c^2}\right) + 20 \cdot \left(\frac{b_c \cdot h_c}{2}\right) = 4.656\ cm^2$
For an entire inner rod ($L_T = 14.7$ cm)	$A_s = 2 \cdot (h \cdot l) + 2 \cdot (h \cdot L_T) + 8 \cdot (d_c \cdot L_T) + 20 \cdot \left(L_T \cdot \sqrt{\left(\frac{b_c}{2}\right)^2 + h_c^2}\right) + 20 \cdot \left(\frac{b_c \cdot h_c}{2}\right) = 57.268\ cm^2$

Table 8 Top view, dimensions and calculated surface areas of the corner rod.

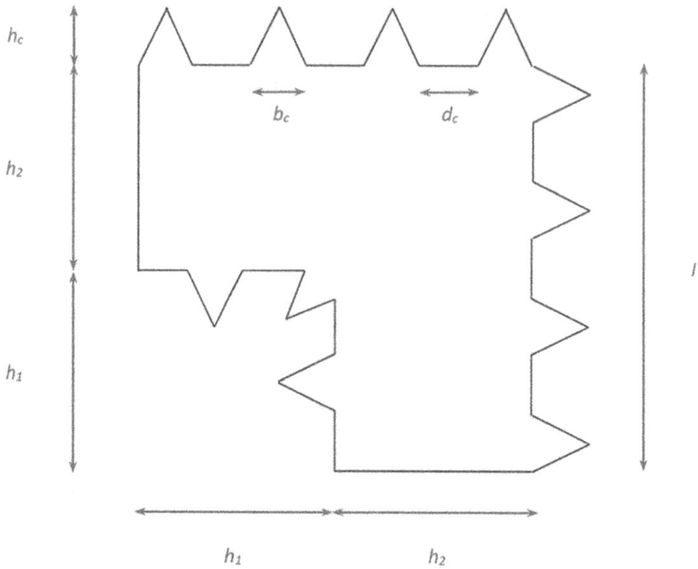

$l = 0.7$ cm
$h_1 = 0.3$ cm
$h_2 = 0.4$ cm
$b_c = 0.1$ cm
$h_c = 0.1$ cm
$d_c = 0.1$ cm

For a corner sample with unit length ($L = 1$ cm)

$$A_s = 2 \cdot (h_1 \cdot h_2 + h_2 \cdot l) + 2 \cdot (h_2 \cdot L) + 10 \cdot (d_c \cdot L) + 22 \cdot \left(L \cdot \sqrt{\left(\frac{b_c}{2}\right)^2 + {h_c}^2}\right) + 22 \cdot \left(\frac{b_c \cdot h_c}{2}\right) = 5.170 \text{ cm}^2$$

For an entire corner rod ($L_T = 14.7$ cm)

$$A_s = 2 \cdot (h_1 \cdot h_2 + h_2 \cdot l) + 2 \cdot (h_2 \cdot L_T) + 10 \cdot (d_c \cdot L_T) + 22 \cdot \left(L_T \cdot \sqrt{\left(\frac{b_c}{2}\right)^2 + {h_c}^2}\right) + 22 \cdot \left(\frac{b_c \cdot h_c}{2}\right) = 63.591 \text{ cm}^2$$

Table 9 Σ and $\ln \Sigma$ values for samples of a freshly unpacked flare.

Sample	Rod	$m(t_0)$ (g)	$\%_{EC}$	A_s (cm^2)	Σ (g cm^{-2})	$\bar{\Sigma}$ (g cm^{-2})	$\ln \Sigma$	$\ln \bar{\Sigma}$
A	Inner	0.7472	2.278	4.656	3.656×10^{-3}	3.750×10^{-3}	−5.611	−5.861
B		0.7937	2.255		3.844×10^{-3}		−5.561	
C	Corner	0.7396	2.268	5.170	3.244×10^{-3}	3.242×10^{-3}	−5.731	−5.732
D		0.7445	2.249		3.238×10^{-3}		−5.733	

For freshly unpacked samples (t_0), the average values of Σ and $\ln \Sigma$ were used.

3.6 Arrhenius equation

The natural logarithm of k_0 and k_1 are graphed as a function of the reciprocal of the temperature, expressed in K^{-1} (Table 11). As shown in Fig. 13 and 14, from the linear regression statistics the dependence of the degradation rate on the temperature is obtained. The opposite of the angular coefficient of the straight line, multiplied by the universal gas constant ($R = 8.314$ J mol^{-1} K^{-1}), returns the activation energy of the process.

Table 10 Σ and $\ln \Sigma$ values for artificially aged samples.

T (°C)	Duration (h)	Sample	Rod	m_i (g)	$\%_{EC}$	A_s (cm^2)	Σ (g cm^{-2})	$\ln \Sigma$
66.5	24	1	Inner	0.7793	2.240	4.656	3.707×10^{-3}	−5.598
	48	2		0.8238	2.227		3.685×10^{-3}	−5.603
	72	3		0.7725	2.204		3.647×10^{-3}	−5.614
	96	4		0.8474	2.177		3.602×10^{-3}	−5.626
71.5	24	1	Corner	0.8137	2.162	5.170	3.103×10^{-3}	−5.775
	48	2		0.7403	2.057		2.952×10^{-3}	−5.825
	72	3		0.7302	2.045		2.935×10^{-3}	−5.831
	96	4		0.7293	2.007		2.881×10^{-3}	−5.850
76.5	24	1	Inner	0.7978	2.190	4.656	3.623×10^{-3}	−5.620
	48	2		0.8099	2.068		3.422×10^{-3}	−5.678
	72	3		0.7169	2.033		3.364×10^{-3}	−5.695
	96	4		0.7995	2.030		3.359×10^{-3}	−5.696
84.5	24	1	Corner	0.7164	2.067	5.170	2.967×10^{-3}	−5.820
	48	2		0.7540	2.008		2.882×10^{-3}	−5.849
	72	3		0.7541	1.944		2.790×10^{-3}	−5.882
	96	4		0.7381	1.913		2.746×10^{-3}	−5.898

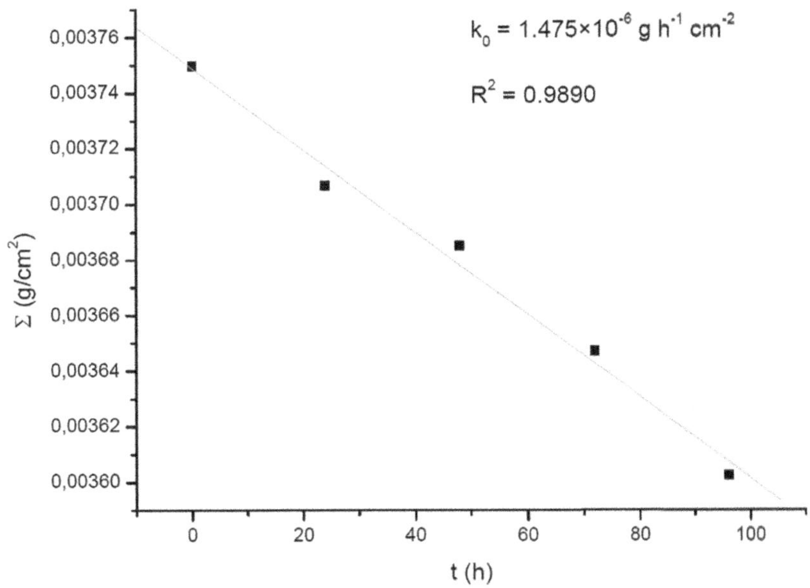

Fig. 5 Graphic evaluation of k_0 at $T = 66.5$ °C.

For the zero-order kinetics (Fig. 13), the linear interpolation is characterized by the following equation, with relative coefficient of determination R^2:

$$\ln k_0 = -7312.46 \cdot \frac{1}{T} + 8.398 \quad R^2 = 0.7207$$

A_0 and Ea_0 are calculated as follows:

$$A_0 = e^{8.938} = 4438.847 \text{ g h}^{-1} \text{ cm}^{-2} = 1.233 \text{ g s}^{-1} \text{ cm}^{-2}$$

$$Ea_0 = -(-7312.45 \cdot R) = 60\,799.19 \text{ J mol}^{-1} = 60.799 \text{ kJ mol}^{-1}$$

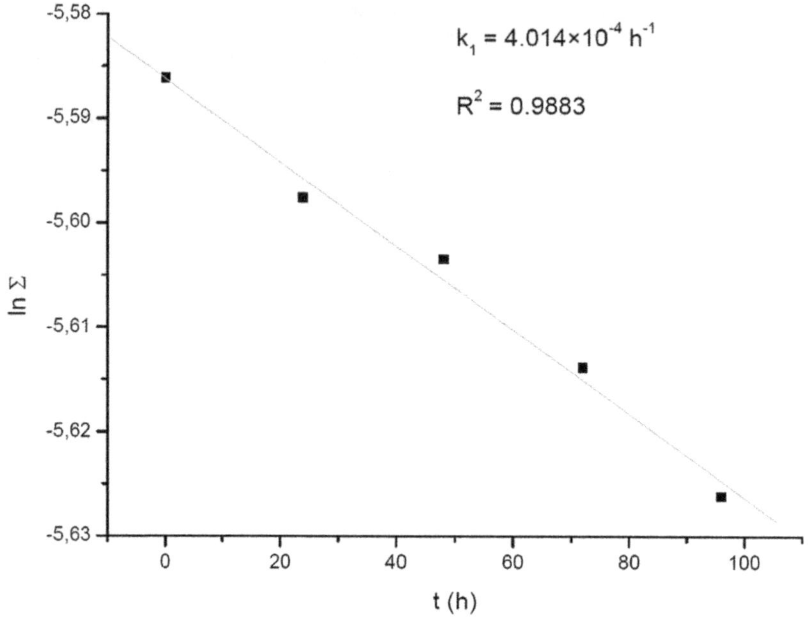

Fig. 6 Graphic evaluation of k_1 at $T = 66.5$ °C.

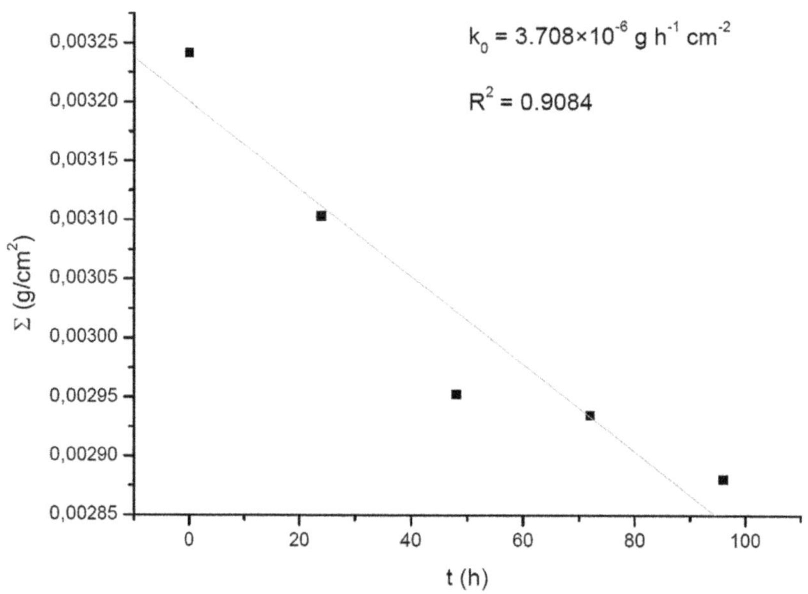

Fig. 7 Graphic evaluation of k_0 at $T = 71.5$ °C.

For the first order kinetics (Fig. 14), the linear interpolation is characterized by the following equation, with relative coefficient of determination R^2:

$$\ln k_1 = -8372.45 \cdot \frac{1}{T} + 17.164 \quad R^2 = 0.7221$$

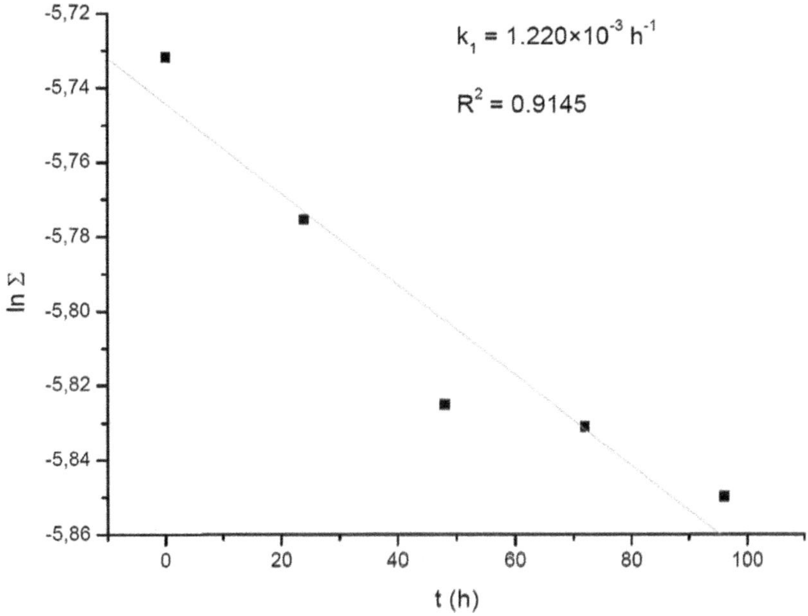

Fig. 8 Graphic evaluation of k_1 at $T = 71.5$ °C.

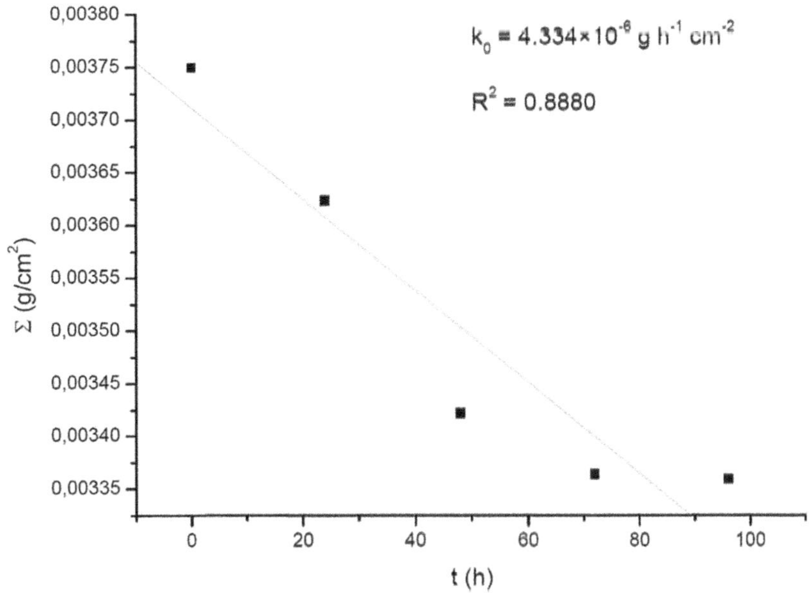

Fig. 9 Graphic evaluation of k_0 at $T = 76.5$ °C.

A_1 and Ea_1 are calculated as follows:

$$A_1 = e^{17.164} = 2.845 \times 10^7\ \text{h}^{-1} = 7.902 \times 10^3\ \text{s}^{-1}$$

$$Ea_1 = -(-8372.45 \cdot R) = 69\,612.45\ \text{J mol}^{-1} = 69.612\ \text{kJ mol}^{-1}$$

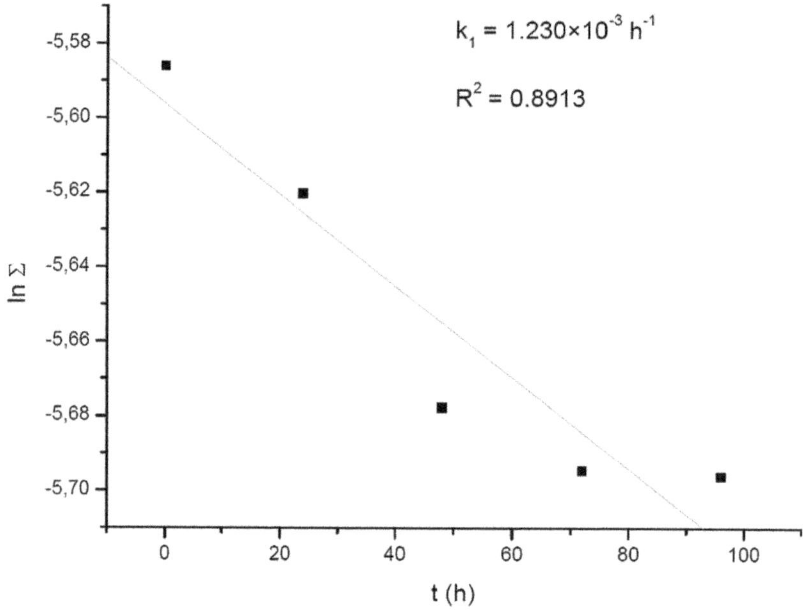

Fig. 10 Graphic evaluation of k_1 at $T = 76.5$ °C.

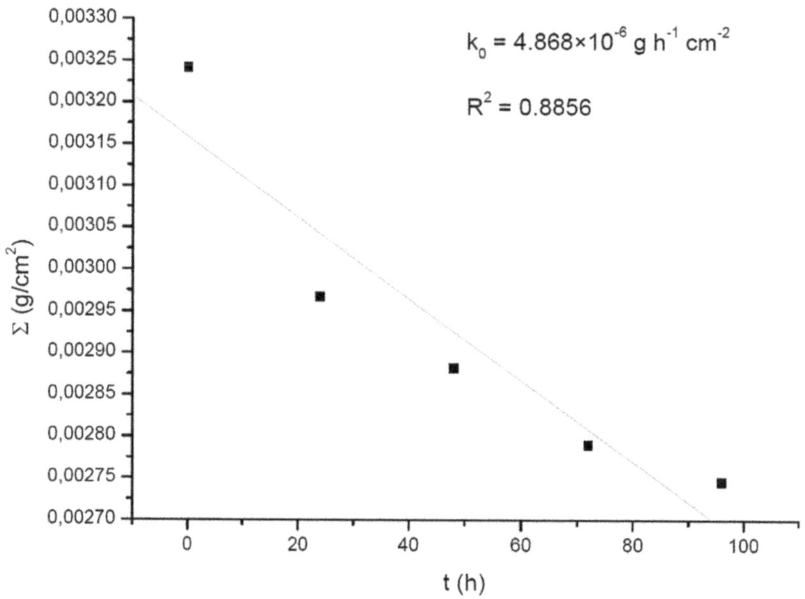

Fig. 11 Graphic evaluation of k_0 at $T = 84.5$ °C.

3.7 Estimation of degradation times in real conditions

Once the kinetic parameters are obtained, the degradation times for the countermeasure in the marine environment can be predicted. In this particular case, the degradation has been predicted for the entire rods (Table 12). In order to calculate the parameter $\Sigma(t_0)$, that will replace $S(t_0)$ in the eqn (6)–(8), the average masses of three central rods and three

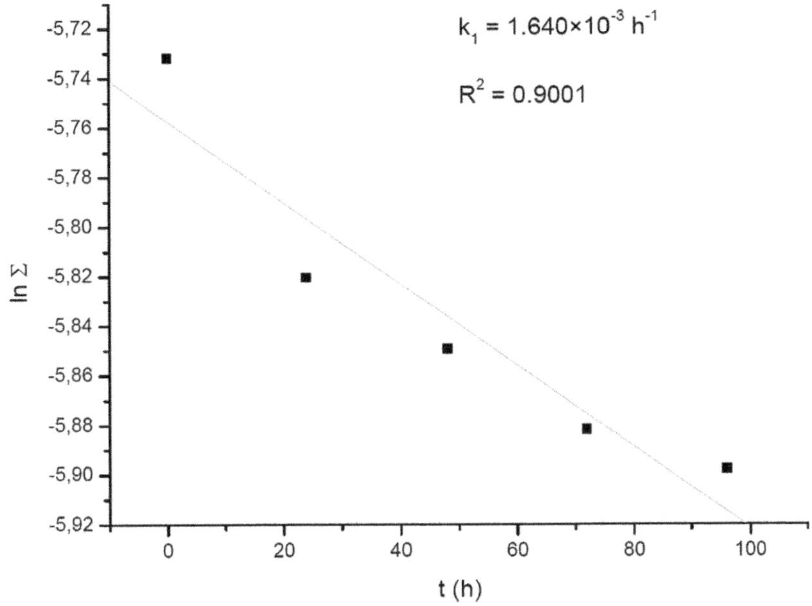

Fig. 12 Graphic evaluation of k_1 at $T = 84.5$ °C.

Table 11 Rate constants as a function of T.

T (K)	$1/T$ (K^{-1})	k_0 ($g\,h^{-1}\,cm^{-2}$)	$\ln k_0$	k_1 (h^{-1})	$\ln k_1$
339.65	0.002944	1.475×10^{-6}	−13.427	4.014×10^{-4}	−7.821
344.65	0.002901	3.708×10^{-6}	−12.505	1.220×10^{-3}	−6.709
349.65	0.002860	4.334×10^{-6}	−12.349	1.230×10^{-3}	−6.701
357.65	0.002796	4.868×10^{-6}	−12.233	1.640×10^{-3}	−6.413

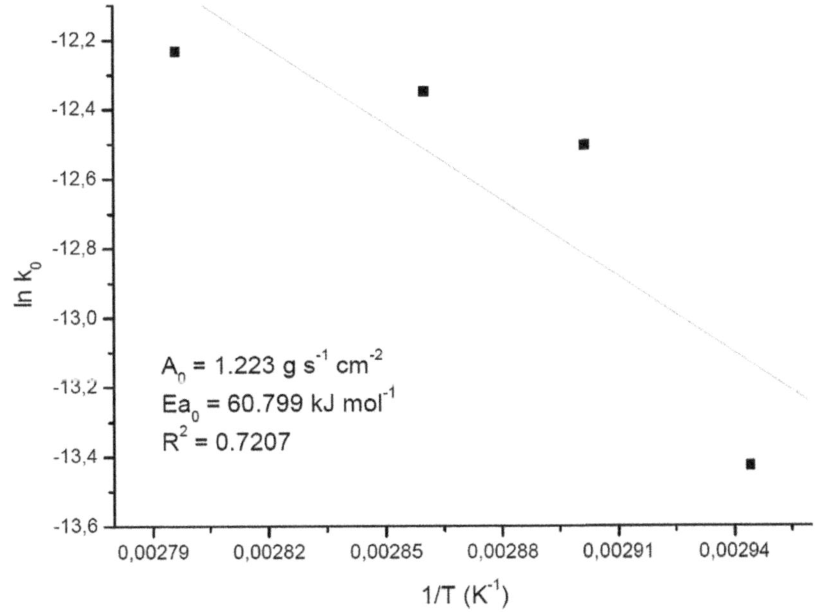

Fig. 13 Graphic evaluation of A_0 and Ea_0.

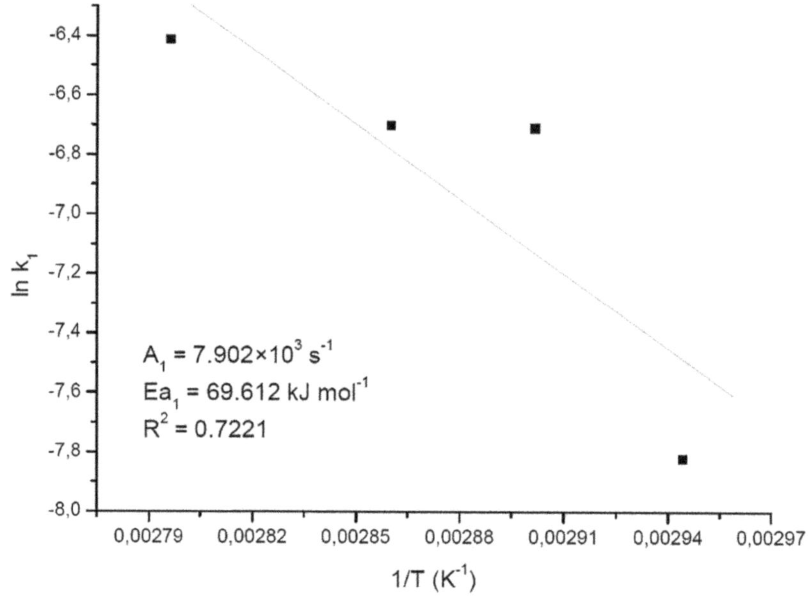

Fig. 14 Graphic evaluation of A_1 and Ea_1.

Table 12 Σ values for entire inner and corner rods.

Rod	$\%_{EC}(t_0)$	$\bar{m}_{rod}$ (g)	L (cm)	A_s (cm^2)	$\Sigma(t_0)$ (g cm^{-2})
Inner	2.266	11.326	14.7	57.268	4.474×10^{-3}
Corner	2.258	10.908		63.591	3.244×10^{-3}

entire lateral rods are used. The length of the entire strip L is expressed in cm.

According to meteorological data reported online,[26] the average annual temperature of the Mediterranean Sea is around 19 °C, equivalent to 292.15 K. By exploiting the Arrhenius eqn (3) and (4) for this temperature, the rate constants equal the following values:

$$k_0 = A_0 \exp\left(-\frac{Ea_0}{RT}\right) = 1.662 \times 10^{-11} \text{ g s}^{-1} \text{ cm}^{-2}$$

$$k_1 = A_1 \exp\left(-\frac{Ea_1}{RT}\right) = 2.830 \times 10^{-9} \text{ s}^{-1}$$

The values of k_0, k_1 and Σ_0 are inserted into eqn (7) and (8) so as to obtain both an estimate of the time necessary for a 80% in reduction and for the total disappearance of the stabiliser:

$$t_{20\%} = \frac{1}{k_1} \ln\left(\frac{1 + \dfrac{k_0}{\Sigma(t_0) \cdot k_1}}{0.2 + \dfrac{k_0}{\Sigma(t_0) \cdot k_1}}\right) = 1.502 \times 10^8 \text{ s} = 4.759 \text{ years}$$

$$t_{0\%} = \frac{1}{k_1} \ln\left(\frac{\Sigma(t_0) \cdot k_1}{k_0} + 1\right) = 2.004 \times 10^8 \text{ s} = 6.349 \text{ years}$$

Inner rod

$$t_{20\%} = \frac{1}{k_1} \ln \left(\frac{1 + \dfrac{k_0}{\Sigma(t_0) \cdot k_1}}{0.2 + \dfrac{k_0}{\Sigma(t_0) \cdot k_1}} \right) = 1.352 \times 10^8 \text{ s} = 4.285 \text{ years}$$

$$t_{0\%} = \frac{1}{k_1} \ln \left(\frac{\Sigma(t_0) \cdot k_1}{k_0} + 1 \right) = 1.790 \times 10^8 \text{ s} = 5.673 \text{ years}$$

Corner rod

To better understand the degradation extent as a function of the seawater temperature, the stabiliser depletion of the entire strips are predicted at 5, 10, 20 and 25 °C: the differences in terms of degradation rate between inner and corner rods (Table 13 and Fig. 15) tend to reduce and disappear as the temperature increases.

In Table 13, the degradation times are reported with the related uncertainties $s(t_{20\%})$ and $s(t_{0\%})$. As can be seen, the relatively low coefficients of determination R^2 (Fig. 13 and 14) generate coefficients of variation higher than 100%. This large variability can be attributed to the pronounced lack of homogeneity in both the geometry of the rods – often characterised by surface irregularities – and the distribution of the stabiliser within the flare pellet. In addition, different samples can show divergent accessibility for the saline solution to reach the embedded nitrate esters and accelerate the degradation process.

3.8 Verification of the prediction provided by the theoretical kinetic model

The validity of the elaborated prediction method is proven by measuring the ethyl centralite percentage in a sample of countermeasure

Table 13 Prediction of the stabiliser depletion as a function of *T*.

			Inner rod		Corner rod	
T (°C)	k_0 ($\text{g s}^{-1}\text{cm}^{-2}$)	k_1 (s^{-1})	$t_{20\%}$ (years)	$t_{0\%}$ (years)	$t_{20\%}$ (years)	$t_{0\%}$ (years)
5	4.716×10^{-12}	6.688×10^{-10}	17.642	23.311	15.808	20.748
10	7.502×10^{-12}	1.138×10^{-9}	10.894	14.443	9.778	12.873
15	1.174×10^{-11}	1.901×10^{-9}	6.835	9.093	6.146	8.115
20	1.810×10^{-11}	3.120×10^{-9}	4.354	5.813	3.921	5.194
25	2.751×10^{-11}	5.037×10^{-9}	2.813	3.770	2.538	3.373

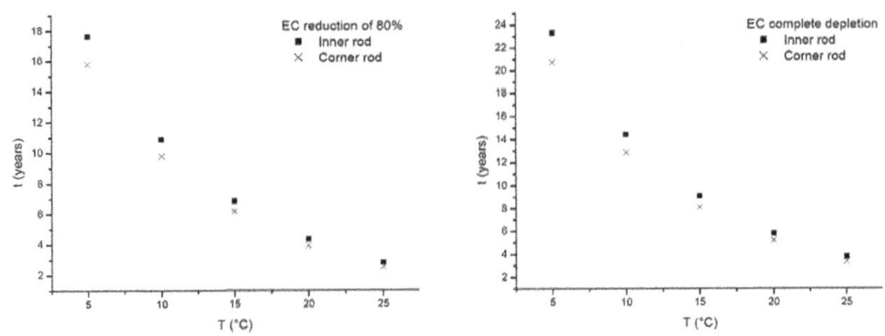

Fig. 15 Depletion time of ethyl centralite as a function of *T*.

Table 14 Quantification of the ethyl centralite percentage in a 334-day naturally aged sample.

Sample	Rod	m (g)	C (mg l^{-1})	V (ml)	m_i (g)	$\%_{EC}$
1	Inner	0.7327	84.007	176.687	0.7705	1.927

kept immersed in sea water, at room temperature, for a period of 334 days.

A sample with a length of 1 cm and a mass of 0.7327 g was taken from one of the inner rods of the countermeasure, dried in a desiccator for 72 hours, prepared as described in par. 2.2 and analyzed by HPLC. The results are reported in Table 14.

Since the initial mass of the analysed sample was not available, the $\%_{EC}$ was calculated using the average mass of three 1-cm long samples from the central rods of a freshly unpacked flare (t_0).

Compared to the $\%_{EC}$ measured at t_0 for the inner rod sample, the remaining fraction of ethyl centralite is equal to

$$\Delta = \frac{1.927}{2.266} \times 100 = 85.006\%$$

equivalent to a decrease of 14 994%, occurred at an average annual temperature of 16 °C. The uncertainty of Δ equals 3.134%.

Under these conditions, the developed kinetic model predicts that the rate constants equal the following values

$$k_0 = A_0 \exp\left(-\frac{Ea_0}{RT}\right) = 1.282 \times 10^{-11} \text{ g s}^{-1} \text{ cm}^{-2}$$

$$k_1 = A_1 \exp\left(-\frac{Ea_1}{RT}\right) = 2.102 \times 10^{-9} \text{ s}^{-1}$$

and that the predicted time to produce the measured decrease is equal to

$$t_{85.006\%} = \frac{1}{k_1} \ln\left(\frac{1 + \frac{k_0}{\Sigma(t_0) \cdot k_1}}{0.85006 + \frac{k_0}{\Sigma(t_0) \cdot k_1}}\right) = 2.796 \times 10^7 \text{ s} = 323.668 \text{ days}$$

with a percentage relative uncertainty that exceeds 100%, because of the low R^2 in the linear regressions of $\ln k_0$ and $\ln k_1$ as a function of $1/T$ (Fig. 13 and 14).

Nevertheless, the difference between the real and the predicted degradation time is only 10 days. In terms of effective reduction, eqn (6) returns the following percentage of stabilizer for a period of 334 days:

$$\Sigma(334\ d) = \left(\Sigma(t_0) + \frac{k_0}{k_1}\right)\exp[-k_1 \cdot t] - \frac{k_0}{k_1} = 3.170 \times 10^{-3} \text{ g cm}^{-2}$$

$$\%_{EC} = \frac{\Sigma(334\ d) \cdot A_s}{m_i} \cdot 100 = \frac{3.170 \times 10^{-3} \cdot 4.656}{0.7705} \times 100 = 1.916\%$$

The calculated percentage is almost comparable to the actually measured value: the two terms differ by only 0.011%.

3.9 Applicability of the NATO AOP-48 kinetic model

In order to evaluate the adequacy of the proposed kinetic model, the n-th order kinetic model described in the AOP-48 publication[23] is directly applied to the measured variation in the EC percentage. By integrating the equation reported in Table 1, it can be seen that the stabilizer percentage as a function of time can be described as follows:

$$S = S_0[1-(1-n)kt]^{\frac{1}{1-n}} = S_0\left[1-(1-n)Ae^{\frac{-Ea}{RT}}t\right]^{\frac{1}{1-n}} \tag{15}$$

To simplify, the $\%_{EC}$ is indicated by S; the term S_0 corresponds to the initial percentage. From the eqn (15) the degradation times, according to an n-th order kinetic model, can be predicted:

$$t = \frac{1}{A}e^{\frac{Ea}{RT}}\left[\frac{1-(S/S_0)^{1-n}}{1-n}\right] \tag{16}$$

The logarithmic form of the eqn (16) provides a linear regression with the relative standard deviation SD:

$$\ln t = \ln\frac{1}{A} + \frac{Ea}{RT} + \ln\frac{1-(S/S_0)^{1-n}}{1-n} \tag{17}$$

$$\mathrm{SD} = 100 \cdot \sqrt{\frac{\sum\left[\ln\frac{1}{A} + \frac{Ea}{RT} + \ln\frac{1-(S/S_0)^{1-n}}{1-n} - \ln(t)\right]^2}{N-3}} \tag{18}$$

The best interpolation of the experimental data is obtained by determining the reaction order n that produces the minimum value of standard deviation SD. In the present case, n values between −1 and 2 are assigned to the reaction order and the relative standard deviation SD is calculated using eqn (18): as reported in Fig. 16, the interpolation of the SD values as a function of n produces a polynomial function with a minimum at a negative value of n (−6.145, as indicated in Fig. 16).

Since the AOP-48 is related exclusively to degradation kinetics with reaction orders between 0 and 2, it can be stated that the kinetic model proposed in the NATO publication is inapplicable to predict the degradation of nitrate esters-based mixtures in seawater.

3.10 Heat flow microcalorimetry

The calorimetric test highlighted – for both the freshly extracted and naturally aged samples – a normalized heat flux far lower than the limit specified in the document of ref. 24.

The energy curves produced by the two types of samples are substantially comparable, except for a slight increase in the heat flow from the aged samples, that can be traced back to the stabiliser depletion (Table 15).

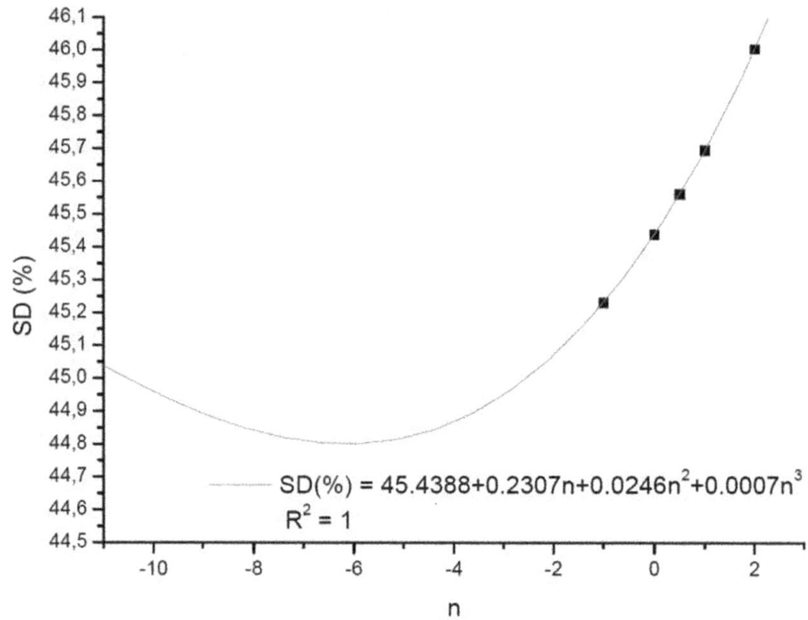

Fig. 16 Dependence of SD (%) on reaction order *n*.

Table 15 Heat flow calorimetry.

Flare	Sample	m (g)	Maximum normalised heat flow ($\mu W\,g^{-1}$)
Freshly unpacked	1	3.0916	18.9
	2	3.1355	19.4
	3	2.9670	17.5
Naturally aged	1	3.1090	24.3
	2	2.7105	54.6
	3	3.1373	61.1

Table 16 Sensitivity ignition test.

Flare	Sample	$T_{ignition}$ (°C)	$\bar{T}_{ignition}$ (°C)
Freshly unpacked	1	153	153.3
	2	153	
	3	154	
Naturally aged	1	157	156.3
	2	156	
	3	156	

3.11 Ignition sensitivity test

The sequence of events during the test shows how, once the trigger temperature has been reached, the development of yellow vapours, associated with the production of nitrogen oxides, was observed. In rapid succession, the reactive phenomena started and propagated. The developed gases pushed the energetic material upwards, resulting in a deflagration outside the test tube. The reaction can be considered complete, given the absence of residues in the test tube and in the immediate surroundings at the end of the process.

The tests led to the results reported in Table 16: the 365-day naturally aged samples have a trigger point only 3 °C higher than the freshly unpacked flares.

4 Conclusions

The comparison of the results from different analytical techniques provides a coherent picture of the degradation process. The decrease rate in the stabilizer concentration is symptomatic of a not particularly rapid decomposition process, which consequently induces a slow inactivation of the investigated countermeasure. The small differences in terms of chemical stability and ignition sensitivity between a freshly disassembled and a 365-day naturally aged countermeasure, prove in real conditions the reliability of the kinetic model for the prediction of the stabiliser depletion. Further replicates of the artificial ageing treatments will surely improve the precision of rate constants and degradation times, although the structural irregularity of the rods, the unbalanced distribution of the stabiliser in the flare pellet and the water accessibility to the reactive sites cannot be adjusted and will remain a consistent source of uncertainties.

The elaborated model allows the estimation, with a certain degree of confidence, of the time needed for the spectral countermeasures to inactivate when they are dispersed in the marine environment. Crucial for the development of the kinetic model was, firstly, to consider the position of prevalence that the hydrolysis mechanism assumes at the typical temperatures of the marine environment and, secondly, to introduce into the kinetic equations the factor As, representative of the influence that the surface of the sample directly exposed to seawater has on the rate of deterioration mechanisms. Assuming that the dumping occurs in Mediterranean waters close to the coast and estimating an average annual temperature close to 19 °C, the model predicts a degradation time for the investigated countermeasures of between five and seven years.

The prediction is confirmed by the results obtained from flare samples immersed in seawater for 334 days: the percentage of stabilizer measured after the immersion period is almost comparable to that estimated using the calculation model, with a residual that stands at 0.011%.

As further confirmation, it has been proven that the slow degradation of nitric esters, although observable and quantifiable by liquid chromatography, do not promptly affect the properties of the dumped countermeasure, which appears, one year later, still chemically stable and susceptible to ignition.

Acknowledgements

This research was funded by the Italian Air Force Aerospace Test Division.

References

1 A. P. Hardt, B. L. Bush and B. T. Neyer, *Pyrotechnics*, Pyrotechnica Publications, Post Falls, 2001.
2 N. Davies, *Pyrotechnics Handbook*, Cranfield University. Defence College of Management and Technology, Shrivenham, 2015.
3 M. Andrews, C. Collet, R. Millar and E. Schlutz, *Proceedings of the 8th International Nitrocellulose Symposium*, MSIAC, Bergerac, Brussels, 2018.
4 E. Roth, V. Weiser, A. Raab and A. Lity, presented at 45th International Annual Conference of ICT, Karlsruhe, 24th–27th June 2014, Pfinztal, Fraunhofer-Institut für Chemische Technologie ICT, 2014.
5 J. Akhavan, *The Chemistry of Explosives*, Cambridge, Royal Society of Chemistry, 2nd edn, 2004.
6 T. Urbansky, *Chemistry and Technology of Explosives*, Pergamon Press, Oxford, 1964.
7 R. Millar, M. Andrews and W. Babcock, *L-221: Lifecycle of Flare Countermeasure. Report 3: Ageing and Degradation*, NATO Headquarters, Brussels, 2018.
8 B. Vogelsanger, *Chimia*, 2004, **58**, 401.
9 M. A. Bohn, N. Eisenreich, M. Doerich and H. Pontius, *Proceedings of 5th International Nitrocellulose Symposium*, Nitrochemie, Spiez, Wimmis, 2012.
10 Lindblom T., PhD thesis, Uppsala University, 2004.
11 *Towards the Monitoring of Dumped Munitions Threat (MODUM)*, ed. J. Beldowski, R. Been and E. K. Turmus, Springer, Dordrecht, 2018.
12 M. I. Greenberg, K. J. Sexton and D. Vearrier, *Clin. Toxicol.*, 2016, **54**(2), 79.
13 *Sea-dumped Chemical Weapons: Aspects, Problems and Solutions*, ed. A. V. Kaffka, Springer, Dordrecht, 2010.
14 M. Heil, K. Wimmer and M. A. Bohn, *Propellants, Explos., Pyrotech.*, 2017, **42**(7), 706.
15 M. A. Bohn, *Proceedings of STO-MP-AVT-268. Utrecht*, NATO Headquarters, Brussels, 2017.
16 Y. Gu, S. Yu, Q. Wang, J. Du and F. Wang, *High Temp. Mater. Processes*, 2022, **41**(1), 589.
17 M. Castagnolo, *Cinetica Chimica*, Lecture for the Bachelor's Degree in Chemistry, University of Bari, 2008.
18 M. Capitelli, R. Celiberto and S. Longo, *Fondamenti di Chimica: Termodinamica e Cinetica Chimica*, Adriatica Editrice, Bari, 2004.
19 M. A. Bohn, in *Ageing Studies and Lifetime Extension of Materials*, ed. L. G. Mallinson, Kluwer Academic/Plenum Publishers, New York, 2001, pp. 449–466.
20 Bohn M. A., *Proceedings of 11th Symp. on "Chemical Problems Connected with the Stability of Explosives"*, Båstad, 24th–28th May 1998, Hansson Listh, Sundbyberg, 1999.
21 M. A. Bohn, *Propellants, Explos., Pyrotech.*, 1994, **19**, 166.
22 M. A. Bohn, presented in part at Atomic Weapons Establishment, Aldermaston, 2007.
23 North Atlantic Treaty Organisation, *AOP-48: Explosives, nitrocellulose-based propellants, stability test procedures and requirements using stabilizer depletion*, NATO Standardisation Agency, Brussels, 2008.
24 North Atlantic Treaty Organisation, *STANAG 4582: Explosive, nitrocellulose based propellants, stability test procedure and requirements using heat flow calorimetry*, NATO Standardisation Agency, Brussels, 2007.
25 G. M. Marion, F. J. Millero, M. F. Camoes, P. Spitzer, R. Feistel and C. T. A. Chen, *Mar. Chem.*, 2011, **126**(1–4), 89.
26 Google LLC, Mediterranean Water Temperature. 2024. Available at: https://www.seatemperature.org/mediterranean-sea#google_vignette (accessed: 25 January 2024).

Computed tomography of a polymer bonded explosive

Robert Cargill

The microstructure of consolidated energetic material formulations influences material properties such as thermal and mechanical, and reactive properties such as shock sensitivity and detonation parameters. Determining the microstructure of consolidated charges without destructive sectioning requires the implementation of non-destructive techniques (NDT) such as X-ray. However, characterisation of microstructure using NDT techniques can be challenging; contrast between ingredients within energetic material formulations can be poor due to similarity in density. NDT can provide both qualitative and quantitative data for assessment, additionally microstructure characterisation data is necessary to develop more accurate material models, particularly at the mesoscale. The Skyscan 1272 micro-Computed Tomography (μCT) was used to image the microstructure of an HMX/HTBP formulation 92/8 wt% pellet. Useful image datasets were obtained achieving a pixel resolution of *circa* 6 μm. Using volume reconstruction software tools metrology, segmentation and porosity analysis was conducted. The results provide useful qualitative information. Although not exactly matching the mass fraction of the formulation, segmentation has been demonstrated.

1 Introduction

The failure mechanisms associated with Polymer Bonded Explosives (PBX's) are well studied and, in principle the concepts are understood. However, the exact nature of failure varies due to the various complexity of the ingredients that make up the composite material. This complexity is summed up by Pietron & Mirkarimi[1] as:

> *"The inherent complexity of PBX composites presents multiple challenges: PBX composites comprise irregularly shaped energetic material powders coated with low weight fractions of polymer, the deposition of which is typically executed via non-equilibrium processes such as slurry coating. Requirements for the polymeric binder are demanding, making binder design and selection challenging. Desirable polymer properties for PBX are often at odds with one another: e.g. good polymer flow characteristics are desirable to enable complete coating of the irregularly-shaped (high explosive) powder surfaces, but good flow often comes at the expense of polymer mechanical stiffness, which is needed to prevent deformation and cracking of PBX under mechanical stress"*

Ignition mechanisms and performance parameters are, to a greater or lesser extent influenced by the microstructure, therefore characterisation of the microstructure is an important part of understanding explosive safety and performance.

Early work into the ignition mechanisms of explosives was undertaken by Bowden and Yoffe[2] they proposed the 'hot spot' theory of ignition.

Ministry of Defence, Abbey Wood, Bristol, UK. E-mail: robert.cargill593@mod.gov.uk

Mechanisms such as friction, pore collapse, shear, fracture, and viscous flow are all influenced by microstructure.

There have been several studies examining the relationship between microstructure and shock sensitivity of explosives[3–5] with morphology, roughness and crystal defects all being important factors. Other explosive performance factors such as critical diameter, detonation wave curvature and corner turning are all effected by microstructure.[6,7]

Stress–strain and mechanical properties relationships of PBX's have also been well studied. Wiegand[8] investigated stress–strain relationships of PBX's and associated failure modes using quasistatic tensile and compression testing, Pietron & Mirkarimi,[1] studied binder properties and microstructure volume growth, Palmer *et al.*[9] and Siviour *et al.*[10] examined the effect crystal particle size on the mechanical strength. Experiments using the Brazilian disk test[9] and split Hopkinson pressure bar[10] have also been widely reported by the University of Cambridge having undertaken many studies. Liu *et al.*[11] and Rae *et al.*[12] used digital image correlation techniques on PBX samples under load to investigate particle displacement.

There have been various studies on Energetic Materials where μCT has been used, Yeager *et al.*[13] used a Xradia Versa 520 Micro XCT to image a variety of PBX formulations and were able to achieve a voxel resolution of *circa* 0.8 μm and achieved accurate segmentation of the material phases. Manner *et al.*[14] looked at *in situ* compression testing of PBX's using a Xradia Carl Zeiss X-ray Microscope operated achieving a voxel size of 5.2 μm. Yuan *et al.*[15] looked at mechanical failure of a TATB based PBX by means of *in situ* Brazilian disk tests achieving resolution of *circa* 20 μm.

The aim of this work is to understand the capability of the SkyScan 1272 and Volume Graphics Software to acquire and analyse 3D volumetric data for microstructure assessment.

1.1 Materials

A curable moulding formulation was prepared by hand mixing a bimodal (Class 2 : Class 3 (1 : 1.85)) 1,3,5,7-tetranitro-1,3,5,7-tetrazocane (HMX) with binder at the ratios listed in Table 1. Butanone (~5 ml) was used as a processing aid to improve the mix quality by thinning the viscous hydroxy-terminated polybutadiene/isophorone diisocyanate (HTPB/IPDI) binder system. The resultant formulation was dried under vacuum to remove residual solvent (−950 mbar, 30 min), and stored in a freezer to retard the curing reaction (−12 °C, 48 h) prior to being pressed.

Details of particle sizes are reported in Table 1. It was noted that the Class 3 HMX presents as a granulated material, agglomerates of which are readily broken apart. Data describing both the particle size distribution of the raw material as received and the corresponding crystallite size distribution revealed by dissociation of agglomerated under the influence of ultrasound, measurements were made on a Mastersizer 3000.

Pressing of the pellet was conducted using a 10 mm diameter press tool, generating a 10 mm right cylinder filled with a loading mass of 1.45 g per pellet. Pellet production details are shown in Table 1. The applied load of *circa* 2000 kg equates to an applied consolidation

Table 1 PBX composition and pressing information.

Compound	Wt%		
HMX Class 3 (coarse)	59.8		
HMX Class 2 (fine)	32.2		
R45-HTLO (HTPB)	7.29		
IPDI	0.71		
Parameter	**Value**		
Hydraulic load	145 bar (approx. 2T)		
Applied pressure	36 220 psi		
Dwell	1 min + 1 min		
Double ended	Yes		
Temperature	Room		
Vacuum	Yes		
Release agent	Lithium stearate top & bottom		
Nature	**Measurement**	**Class 3 diameter (μm)**	**Class 2 diameter (μm)**
Stirring/agglomerates	Dx (10)	82 ± 2	1.21 ± 0.9
	Dx (50)	290 ± 5	40 ± 10
	Dx (90)	700 ± 50	220 ± 130
Ultrasound/crystallites	Dx (10)	3.35 ± 0.04	0.6310 ± 0.0001
	Dx (50)	22.1 ± 0.2	2.030 ± 0.003
	Dx (90)	46.8 ± 0.8	11.60 ± 0.08

pressure of the moulding powder of 36 220 psi, this is of the same magnitude that would be typical of service use pressings.

Due to the reactivity of IPDI and water, it was decided that density measurements would be made following cure of the material at 70 °C, 7 days.

1.2 Methodology

Micro computed tomography was undertaken on a Bruker Skyscan 1272 X-ray microscope equipped with a 16 Megapixel Complementary metal-oxide semiconductor (CMOS) detector. The X-ray source can operate up to 100 kV (20–100 kV) at a maximum of 10 watts. The equipment can accommodate samples up to 70 mm × 70 mm and achieve resolutions down to 0.45 μm.

The sample was mounted on the sample stand using a foam adhesive pad. The sample was imaged using at a variety scanning and reconstruction parameters to optimise the image quality in terms of phase contrast and artefact minimisation. The parameters selected for image acquisition and reconstruction are given in Table 2.

1.3 Volume graphics analysis

The following volume graphics software tools were used:

1. VGEasyPore – An algorithm to identify voxels as indications based on local grey values in relation to a specified contrast threshold.
2. VG surface determination – A segmentation method whereby material boundaries are determined using the isovalue approach.
3. Greyscale thresholding – Manual selection of grey value thresholds from the image histogram for segmentation.

Table 2 CT scan and reconstruction parameters.

Scanning		Reconstruction	
Voltage (kV)	60	Number of slices	1924
Power (μA)	166	Angular step	0.2
Pixels (row × column)	2048 × 0248	Smoothing	3 (Gaussian)
Image pixel size	6	Ring artefact correction	30
Filter	0.25 mm Al	Beam hardening correction	33%
360° scanning	Yes 0.2° step		
Frame averaging	7		
Scan time	3 h 16 m 26 s		

Table 3 Pellet physical dimensions.

	Software	Calliper	Archimedes
Height (mm)	10.62	10.78	n/a
Diameter (mm)	9.9	10.07	n/a
Mass (g)	n/a	n/a	1.45
Volume (mm^3)	802.58	817.49	838.15

4. VG region growing tool – After setting a seed point a flooding algorithm is used to expand the selection to new voxels within the specified tolerance bands.

2 Results

The density of the pellet was measured by Archimedes method at 1.73 $g\,cm^{-3}$. The theoretical maximum density (TMD) calculated from literature values for HTPB, HMX and IPDI is 1.83 $g\,cm^{-3}$ making the pellet 94.5% TMD. Pellet dimensions were measured by calliper prior to scanning. Pellet volume can be calculated from calliper dimensions however the pellet was not a perfect cylinder, there was some deformation when being retrieved from the press tooling. Pellet volume could also be estimated from the Archimedes density measurement and mass, however only the loading mass was measured and mass loss during pellet cure and drying is likely. Measured physical pellet properties are compared to those obtained by the software at Table 3.

The volumetric data was visualised in the software and subject to qualitative assessment, and quantitative assessment using a the tools within the volume graphics software package.

The visualised volume shows clear greyscale contract between phases as shown in Fig. 1 and 2. Voids and porosity can be seen in the reconstructed data throughout the pellet volume (see Fig. 3). There are many observations of extended cracks present in the volume, often these are, as expected located along particle boundaries where packing geometries become increasingly dominant. There were no indications of high-density inclusions.

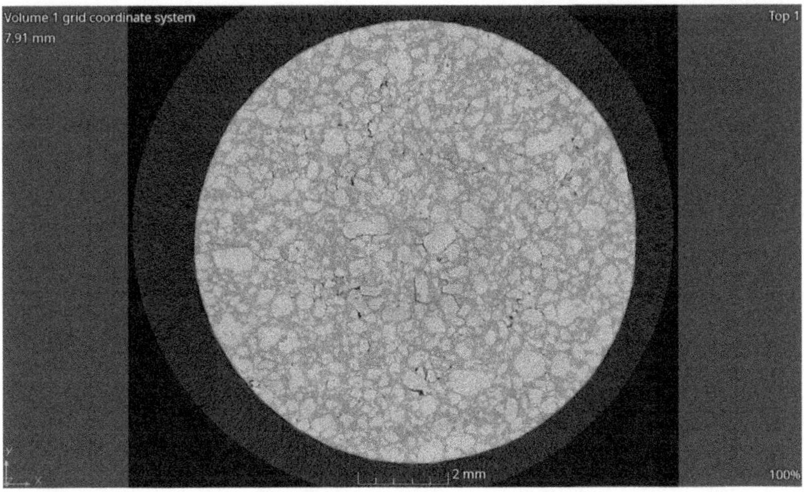

Fig. 1 Axial plane slice.

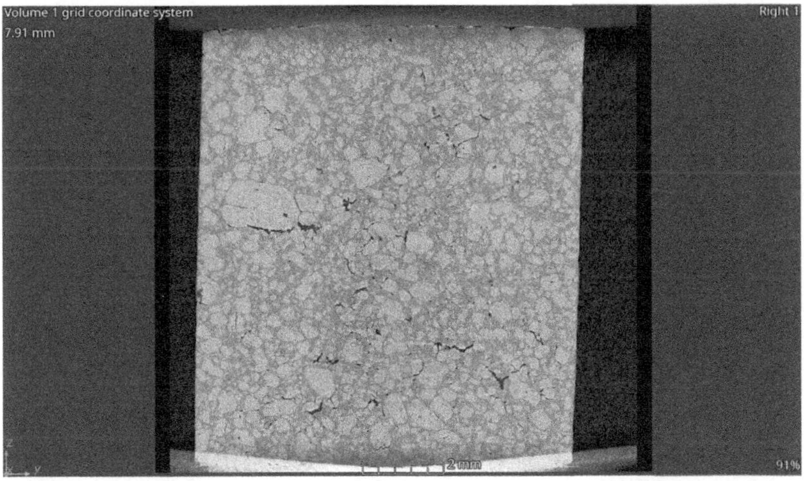

Fig. 2 Longitudinal plane slice.

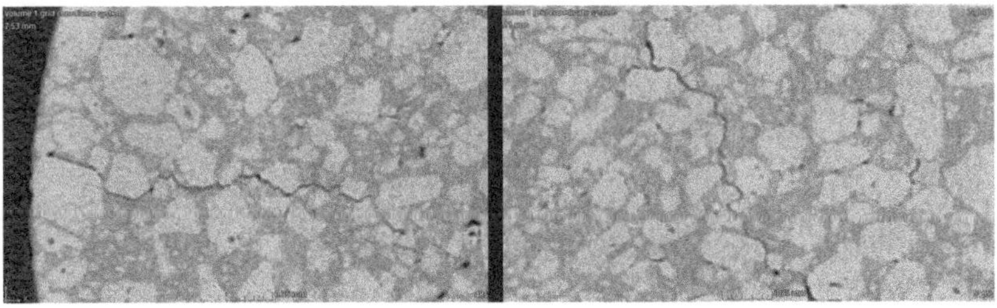

Fig. 3 Example of particle boundary 'cracks'.

A surface determination was applied to the pellet using a combination of manual greyscale thresholding and the region growing tool to separate the pellet from the foam pad.

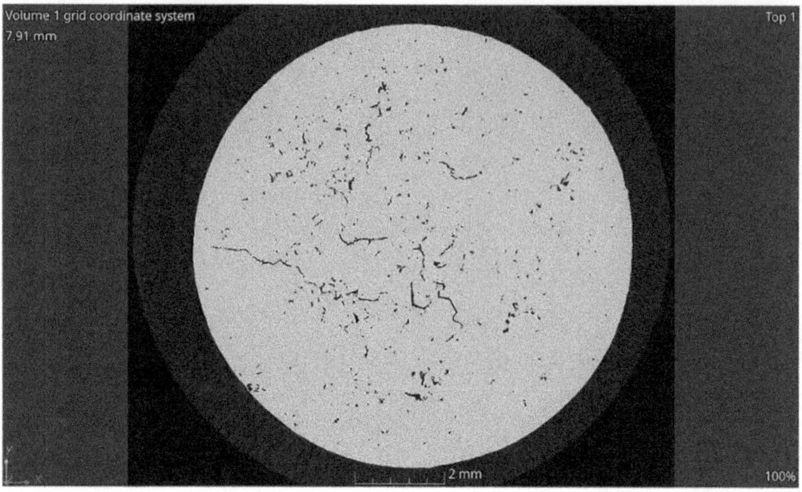

Fig. 4 Porosity analysis axial plane slice.

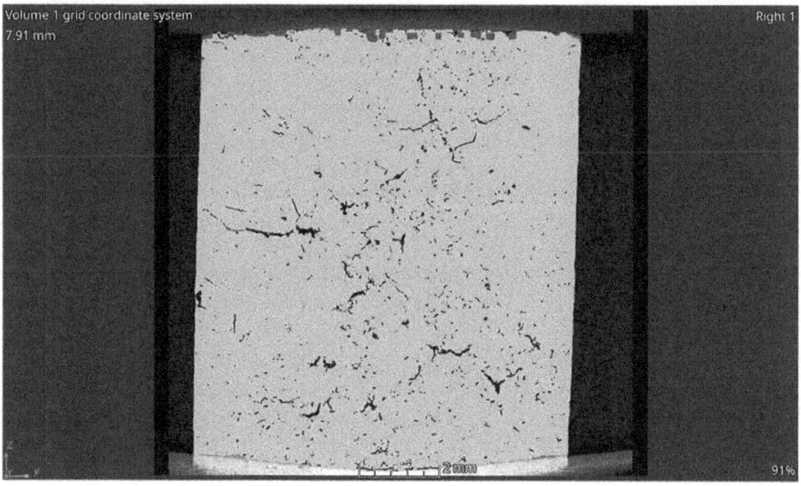

Fig. 5 Porosity analysis longitudinal plane.

Pellet dimensions and volume was calculated in the software by voxel count within the determined surface.

Porosity analysis was undertaken using the VGEasyPore tool set at a 25% threshold value. The software algorithm returned over 85 000 individual pores. The pellet porosity from voxel count was calculated at 17.66 mm^3 or 2.20% (see Figs. 4 and 5).

A random region was segmented using the surface determination isovalue algorithm to sperate the HMX and binder phases as shown in Fig. 6. Based on voxel count the mass fraction of HMX was calculated at approx. 71% representing an underestimation. Metrology of some of the individual particles shows the smallest particles detected being in the order of 40 µm, the particles below this size are likely dispersed in the binder and cannot be resolved.

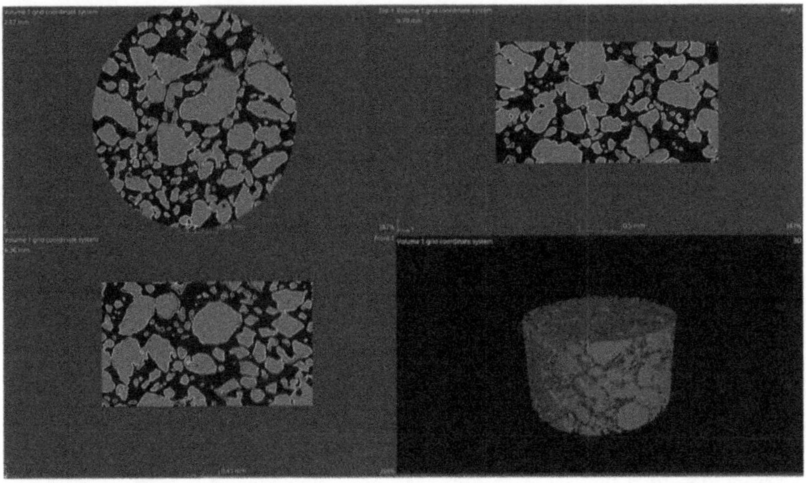

Fig. 6 Segmentation of pellet subsection by grey value (axial, longitudinal, frontal and 3D – top to bottom left to right).

The obtained pixel resolution for this dataset was 6 μm, allowing 3 pixels for edge determination the approximate spatial resolution is 20 μm. The finer particles have almost certainly not been resolved. Particles in the order of 40 μm would only be 7 pixels across, with finer particles being even fewer pixels, such few pixels make determination of boundaries challenging coupled with gradual grey scale changes at boundary make it difficult for the software to determine thresholding and apply edges with accuracy.

3 Conclusions

Understanding energetic materials is essential for their safe operation, therefore the ability to characterise their functional, safety, mechanical and performance properties is vital. A service representative PBX was imaged using micro X-ray computed tomography.

Volume Graphics software and its software algorithms were used for dimensioning, porosity analysis and segmentation of phases. Software measurements were within a 2–5% range of the physical measurements taken for diameter, height, and volume.

Segmentation of phases has been demonstrated with a software calculated mass fraction of 71% against an actual mass fraction of 92%.

Further work is required to optimise both image acquisition and software analysis, particularly in relation to phase segmentation.

Acknowledgements

The author wishes to acknowledge Dr Christopher Stennett, Dr Jack Brash and Mr Matthew Goldsmith from the University of Cranfield for their support.

References

1 J. J. Pietron and P. B. Mirkarimi, *Propellants, Explos., Pyrotech.*, 2022, **47**, e202100379.

2 F. P. Bowden and A. D. Yoffe, *Initiation and Growth of Explosion in Liquids and Solids*, Cambridge University Press, 1952, reissued 1958.
3 B. Khasainov, B. Ermolaev, H. N. Presles and P. Vidal, *Shock Waves*, 1997, **7**, 89.
4 V. J. Bellitto and M. I. Melnik, *Appl. Surf. Sci.*, 2010, **256**, 3478.
5 H. Czerski and W. G. Proud, *J. Appl. Phys.*, 2007, **102**, 113515.
6 K. Parfenov and I. M. Voskoboinikov, *Combust., Explos. Shock Waves*, 1969, **5**, 240.
7 J. W. Forbes, E. R. Lemar, G. T. Sutherland and R. N. Bake, Detonation Wave Curvature, Corner Turning and Unreacted Hugoniot Of PBXN-111, Naval Surface Warfare Centre, 1992.
8 D. A. Wiegand, *Proceedings of the Tenth American Physical Society Topical Conference on Shock Compression of Condensed Matter*, American Institute of Physics, 1998.
9 S. J. P. Palmer, J. Field and J. M. Huntley, *Proc. R. Soc. A*, 1993, **440**, 1909.
10 C. R. Siviour, M. J. Gifford, S. M. Walley, W. G. Proud and J. E. Field, *J. Mater. Sci.*, 2004, **39**, 1255.
11 Z. Liu, H. Xie, K. Li, P. Chen and F. Huang, *Polym. Test.*, 2009, **28**, 627.
12 P. Rae, H. T. Goldrein, S. Palmer, J. Field, R. White and A. L. Lewis, The use of digital image cross-correlation (DICC) to study the mechanical properties of a polymer bonded explosive (PBX), International Detonation Symposium, 2002.
13 J. D. Yeager, L. A. Kuettner, A. L. Duque, L. G. Hill and B. M. Patterson, *Materials*, 2020, **13**, 4517.
14 V. W. Manner, J. D. Yeager, B. M. Patterson, D. J. Walters, J. A. Stull, N. L. Cordes, D. J. Luscher, K. C. Henderson, A. M. Schmalzer and B. C. Tappan, *Materials*, 2017, **10**, 638.
15 Z.-N. Yuan, H. Chen, J.-M. Li, B. Dai and W.-B. Zhang, *Materials*, 2018, **11**, 732.

The effects of surface finish on the signal return for Het-V experiments using copper†

Joseph Henry Nelhams

The surface finish of a material plays a key role in the performance of optical diagnostics such as Heterodyne Velocimetry (Het-V). Without an appropriate surface finish being applied to the target surface, little to no data may be collected. This study investigates the variances in signal return over a range of angles between five different surface finishes applied to copper to optimise the design of experiments utilising these types of diagnostics. The findings of this paper show that all tested surface finishes provided a high enough signal return at angles up to 45° to be used within experiments at AWE.

1 Introduction

Surface reflectance plays a critical role in the performance of Heterodyne Velocimetry (Het-V), a laser-based diagnostic used in shock experiments to measure the velocity of dynamic surfaces.[1,2] Specular surfaces provide high signal return levels once aligned, but most of the return signal is lost when misaligned. Diffuse surfaces provide lower signal returns when aligned, but the signal return levels during misalignment are higher than that of specular surfaces allowing data collection at wider angles. Experiments using Het-V require the target surface to be roughened to create a diffuse return signal, which allows data collection during dynamic misalignment.

One of the experiments commonly performed at AWE, which utilises the Het-V diagnostic, is the cylinder test. A copper tube is packed with explosive and detonated at one end face, this causes a detonation wave to travel from one end of the explosive cylinder to the other. As the detonation wave travels along the explosive cylinder, it causes a shock wave to form in the surrounding copper tube. As this shock wave travels though the copper tube to the outer wall, it causes the outer copper tube wall to expand.[3,4] Het-V probes measure the expansion of the copper tube wall. The tube wall expands with a time-dependant angle towards the Het-V probes, and throughout the duration of the experiment, the point at which the probes is measuring gets increasingly misaligned from the initial static alignment. Therefore, portions of the surface of the cylinder are roughened from the initial condition of R_a 0.8 μm to create a diffuse reflectance, which ensures a signal return to the probe throughout the duration of the experiment. However, a surface which is too rough may not return a signal strong enough to be usable. For the class 1 frequency shifted Het-V system used on cylinder tests at AWE, the acceptable signal

AWE – Nuclear Security Technologies, UK. E-mail: Joseph.Nelhams@awe.co.uk

loss from the surface is up to −55 dB beyond this point the return signal would not trigger the signal amplifiers.

Currently, there are several methods to create a roughened surfaces and there is no quantifiable evidence supporting their use. Therefore, this study has investigated five common surface roughening techniques to identify the optimal method(s).

2 Method

Five copper tokens were prepared with different surface finishes. Copper was the selected material due to it being commonly used in hydrodynamic experiments, such as cylinder tests.

2.1 Surface preparation

Prior to roughing, each copper token was lapped with a 40 μm diamond suspension to provide a common starting flatness, after which each tokens flatness was measured using an optical flat. The copper tokens were lapped to a flatness of less than 1.5 μm before the final finish was applied.‡

The surface finishes selected were as follows: hand roughened using 30 μm grit lapping film (30 μm), hand roughened using 6 μm grit lapping film (6 μm), hand roughened using 30 μm then 6 μm grit lapping film (Mix), a machined surface finish where the copper token would be skimmed on a lathe (Machined). Additionally, one lapped token that had no additional finish applied was tested (Lapped). Images of these tokens are shown in Fig. 1.

2.2 Data collection

The copper tokens were positioned on the central point of rotation of the multi-axis gimbal, which could be set to a specified angle between −90° and +90° with an accuracy of 0.01°. A Het-V probe was mounted on a gimbal and rotated around the copper token, so that measurements were taken across the surface as the probe rotated around the centre of the target. This ensured that the measurements would not be taken from a single point that could be an outlier in relation to the entire surface; instead, measurements were taken across the surface which is more indicative to what occurs during a dynamic experiment. This setup is shown in Fig. 2.

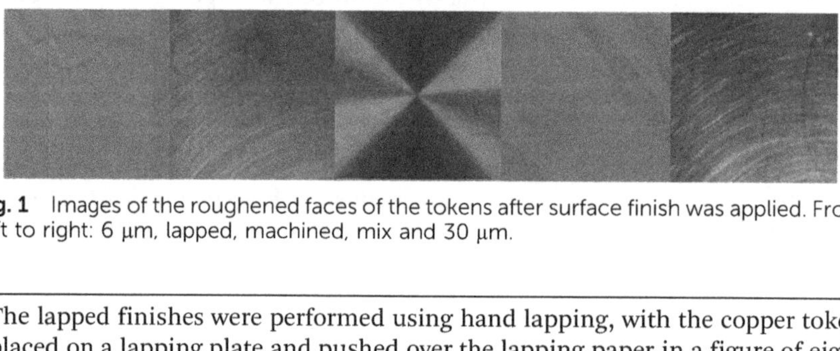

Fig. 1 Images of the roughened faces of the tokens after surface finish was applied. From left to right: 6 μm, lapped, machined, mix and 30 μm.

‡The lapped finishes were performed using hand lapping, with the copper token placed on a lapping plate and pushed over the lapping paper in a figure of eight motion until a uniform roughness had been achieved across the surface.

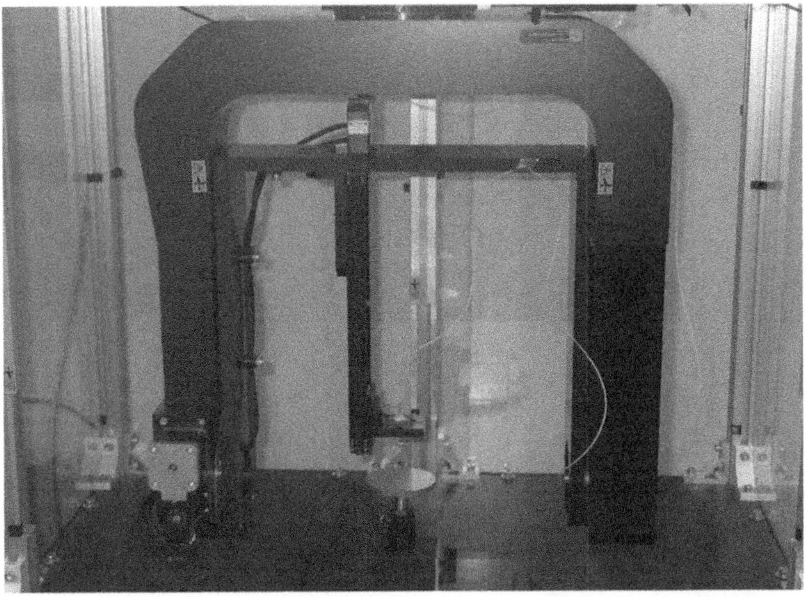

Fig. 2 An image of the experimental setup showing a copper token positioned at the central point of rotation of the 3-axis gimbal. To the left outside the image was the OBR, and the computer which controlled the entire setup.

The HetV probe on the gimbal was connected to an optical backscatter reflectometer (OBR), and mounted onto the translation stage of the gimbal, with the laser spot aligned to the centre of the copper token. The OBR provides surface loss measurements in dB with a measurement uncertainty of ±0.825 dB. The back reflectance was measured for each token within a range of angles varying between -2° and $+45^\circ$, with the majority of measurements between -2° and $+15^\circ$ as this is the range a cylinder test commonly measures.

Both the gimbal and the OBR were controlled using a LabVIEW script; a loop was setup which controlled the gimbal to stop at 127 different points across a range from -2° to $+45^\circ$ and after the gimbal reached the programmed position, the loop would wait for 1 minute before taking a measurement from the OBR. The pause in between points was introduced to minimise the vibrations caused by the gimbal, which could introduce unwanted noise into the data. After one loop the setup was realigned, and a repeat of the measurements was performed.

3 Results

Each dataset was normalised so that the measured scatter of the fibre to the probe was identical between measurements. The average of the two surface measurements of each surface finish was then plotted on a scatter chart. This is shown in Fig. 3 with the data smoothed using a 4-point moving average to highlight trends in the data.

The results show that each surface provides a similar signal return after correction, with signal returns at 0° being between -15 dB and -28 dB,

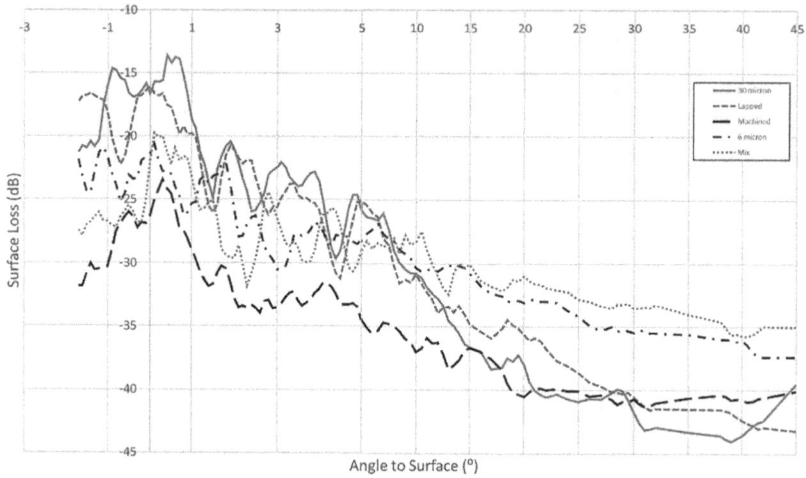

Fig. 3 Scatter graph showing the mean data from all surface finishes plotted with 4-point moving average smoothing. Note that the *x*-axis uses a higher resolution between −3 and +5° to better present the higher density of data in this region.

which is considered a strong signal return with regards to the Het-V system used. The rougher surfaces maintain a stronger signal return across all angles, when compared against the smoother surfaces such as a lapped or machined finish. Notably, all surfaces provided a return signal above −45 dB at all measured angles, so all surfaces are suitable to use with the Het-V system considered.

4 Conclusions

Five copper tokens with different surface finishes had their surface return measured from −2° to 45° to investigate which surface was best for shock experiments that measure the velocity of dynamic surfaces. The data suggests that all surface finishes would provide usable data up to 45° with a class 1 frequency shifted Het-V system. The rougher surfaces 6 μm and Mix have the smallest signal loss difference between 0 and +45°, making them more attractive for any experiments where dynamic misalignment of the probe to the surface is expected. At shallow angles (between −2 to +2°) the smoother surfaces provided less signal loss, this is as expected as the light return is more specular rather than diffuse.

References

1 W. Mullins, *et al.*, *Opt. Eng.*, 2022, **61**(6), 065102.

2 O. T. Stand, *et al.*, *26th International Congress on High-Speed Photography and Photonics*, 2005, p. 5580.

3 B. D. Sutton, J. W. Ferguson and A. N. Hodgson, *AIP Conf. Proc.*, 2017, **1793**, 030032.

4 A. N. Hodgson, J. W. Ferguson and B. D. Sutton, *AIP Conf. Proc.*, 2017, **1793**, 160006.

Section 2
Artificial Intelligence, Simulation and Modelling

Prediction of material properties of energetic materials using machine learning methods

Jan Langer* and Moritz Heil

The application of modern machine learning methods provides new possibilities for gaining information from existing data in many areas. In the case of energetic materials, such methods can help to gain a deeper understanding of the interaction of various influencing factors through exploratory data analysis, both for existing formulations and for the development of new ones. The prediction of specific material properties is also conceivable. The prerequisite for the application of machine learning methods is a set of high qualitative and quantitative data. For this purpose, a new extended data set was created based on an existing data set from the Fraunhofer ICT (Institute for Chemical Technology) Thermodynamics Database in combination with data retrieved from the public PubChem database. Several models were developed with this data set to predict the enthalpy of formation based on information about the composition of substances. Different methods were used for model development, ranging from simpler methods such as multivariate linear regression and partial least squares to more complex methods such as ridge regression and random forest, and even to deep neural networks. The following work describes an iteration of the development cycle of these models, starting with data collection and preprocessing, continuing with model building and hyperparameter tuning, and finally, evaluating and assessing the resulting models.

1 Introduction

The classical approach to developing new formulations of energetic materials from design, synthesis, characterization to experimental verification and evaluation is a very complex, lengthy, and costly process. To get an idea of the properties of the final product before going through this entire process, some of them, such as the enthalpy of formation (EOF), can be predicted after the design phase using various simulation methods. These methods range from the macroscopic level to the quantum mechanical level.[1,2] The latter enables reliable predictions, but at the expense of lengthy simulation times,[3] especially for more complex molecular structures, which can easily take several days to weeks.[4]

The rapid developments in the field of artificial intelligence (AI) and machine learning (ML) over the last ten years have now opened new possibilities for using these methods to predict the material properties of energetic materials.[5–7] Such an adapted model is then able to make predictions about the target parameters within a few seconds,[8] regardless of the complexity of the structure.[9] The accuracy of the prediction depends on the selected ML method, the structure or hyperparameters and the training of the model, which is largely determined by the underlying data set.[10] The basic prerequisite for a good ML model is therefore a qualitatively and, if possible, quantitatively high-quality data set with

Fraunhofer Institute for Chemical Technology, Joseph-von-Fraunhofer-Str. 7, 76327 Pfinztal, Germany. E-mail: info@ict.fraunhofer.de

which the models can be trained. The quantity of this data is often difficult to achieve in the context of working with chemical formulations because, as already mentioned, these are usually only generated by complex and expensive developments. Fortunately, from decades of research on energetic materials at the Fraunhofer ICT, we have a data set available that has been used to create and evaluate various ML models. The goal of this work was not only to develop an ML model for predicting the standard EOF of molecules, but also to gather experience on the extent to which these methods are suitable for transferring them to other issues such as the prediction of mechanical/physical properties.

There have been a few efforts to develop models for predicting the EOF,[11–15] but in most cases the training data set for these models only contains a few hundred well-curated and selected data points, usually from literature sources. With a data-driven approach, as is common with the current state of the art developments of these models, this approach leads to good results in the scoring of the models but can also lead to poor generalization of the models. In this approach, a significantly larger data set was used, which was not analyzed in depth during compilation and thus contains a not to be underestimated number of incorrect entries, which negatively influence the accuracy of the models, but also lead to greater robustness of the models and at the same time counteract early overfitting during training.

2 The data set

The data set used to train the models is fundamentally based on the so-called ICT Thermodynamics Database (TdDb) and comprises around 11 500 entries of different molecules.[16] Each entry contains as descriptors the name and/or CAS (Chemical Abstracts Service) number, a classification, type and number of elements contained, density and the standard EOF of the molecule. In some cases, the name of the molecule is only available in German, only in English or in both languages. The classification is a description of the use of the molecule such as energetic filler, oxidizer, stabilizer or similar. The descriptors listed here are not sufficient to enable a model to understand the correlation between substance and the target variable, the standard EOF. In this chapter we will showcase the road to building a useful data set for model training purposes and which problems may occur on the way.

As a first step to expand the data set with additional information, an attempt was made to obtain the SMILES[17] (Simplified molecular input line entry system) codes for the individual entries *via* automated queries to the public PubChem[18] database using a REST (representational state transfer[19]) API (application programming interface) (see Fig. 1). The SMILES codes are a common system that allows representation of the molecular structure of any molecule by a simple string of characters. The automatic query was successful in about 83% of the cases, so that in the end the corresponding SMILES code was available for 8721 entries

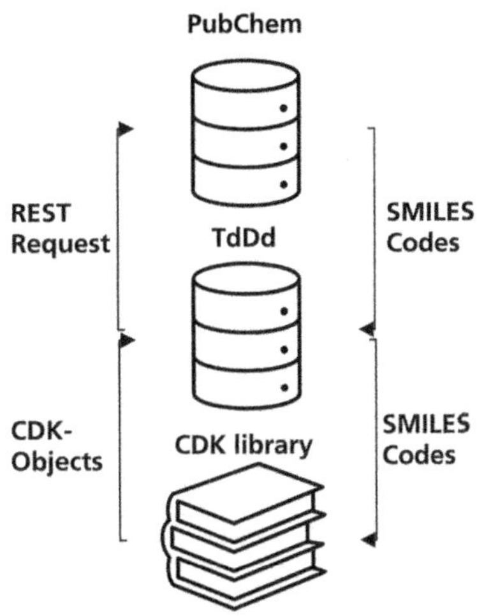

Fig. 1 Information gathering *via* PubChem database and CDK library.

in the TdDb. The two main reasons why such a query failed were either that the molecule does not exist in the PubChem database or that only the German substance name was available as an identifier. In future, the latter case is to be solved by an automatic translation of the German names using a DeepL[20] or FhGenie[21] API. However, even if the query is successful, this does not guarantee that the SMILES code obtained is the correct one for the molecule. To avoid incorrect entries in the training data set, several verification methods have been developed, including, for example, a comparison of the sum formula or a comparison of the expected molecular weight. After this verification, the number of successful queries is further reduced to 6313 entries, which is now only 60% of the original data set.

With the help of the SMILES codes obtained, detailed descriptions of a molecule can now be generated using corresponding cheminformatics libraries. In this case, the Chemistry Development Kit[22] (CDK) was used, which represents the molecules in an object-oriented nested format as JAVA objects. Further descriptors for the training data set can now be obtained from the CDK objects. First, we look at the information on the bonds contained in a molecule, namely the bond type and number of bonds per molecule which can be extracted from the CDK objects. The original data set contains a total of 507 different bond types, whereby many of these bonds are only very rarely represented and thus lead to a very low variance of the corresponding variables. Fig. 2 shows an overview of the distribution of the number of bonds. To avoid this problem, it was decided to initially use only molecules consisting of carbon, nitrogen, oxygen and hydrogen elements as training data. This reduced the number of different bond types to only 14.

In addition to the restriction of the molecules used based on the contained elements, a further restriction was made based on the target parameter space of the standard EOF. The standard EOF of most of the entries in the TdDb lies in a value range between $-1000\ \text{kJ mol}^{-1}$ and

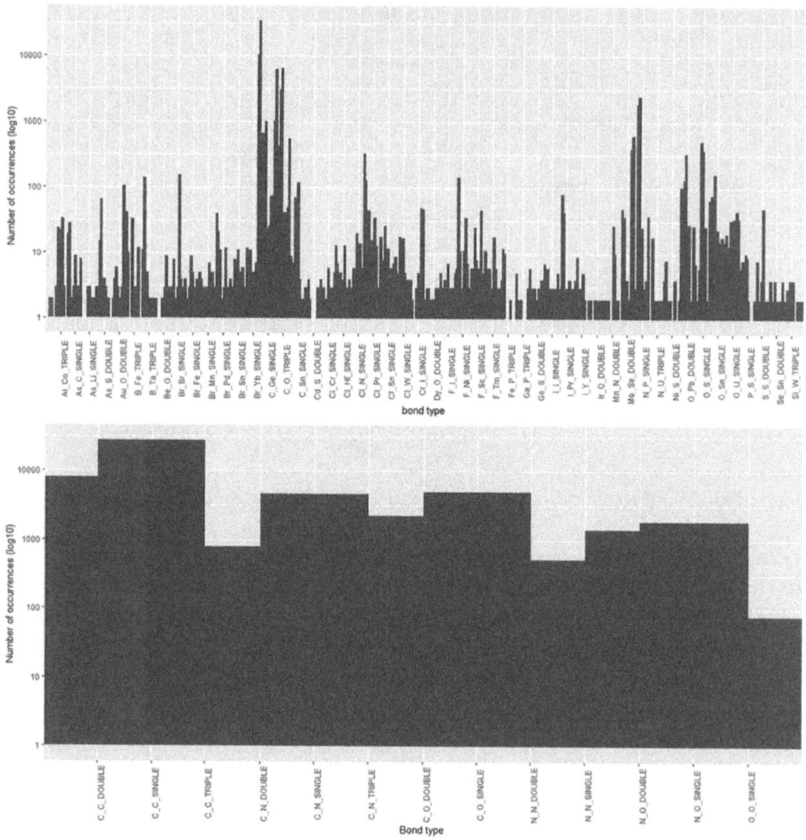

Fig. 2 Distribution of the different bond types. This first histogram includes all 504 different bond types, for better readability only every 10th bond type is labelled on the *x*-axis. The second histogram shows the remaining bond types and the number of their occurrences after limiting the molecules to C, N, H, and O elements.

1000 kJ mol^{-1} as shown in Fig. 3. Outliers outside this value range cause problems when training the model due to their underrepresentation in the data set. Therefore, all entries with a standard EOF outside this value range were removed from the data set. These restrictions on the data set reduce the number of entries to 4295.

In addition to the information about the bonds of a molecule, further descriptors can be derived from the CDK objects, such as the aromaticity and the oxygen balance. The aromaticity only indicates in the form of a Boolean variable whether the molecule contains at least one aromatic ring or not. From the TdDb, the total number of atoms contained, as well as which and how many elements are contained, could already be obtained as descriptors. The total number of descriptors used to develop the first models was therefore 21. The best model achieved an average absolute error of approx. 90 kJ mol^{-1} when making predictions on a test data set.[23] The data set was adapted again and extended by further descriptors within the next iteration of the cycle of data preprocessing, modelling, and model training, by adding the so-called fingerprints, a code that represents the structure of a

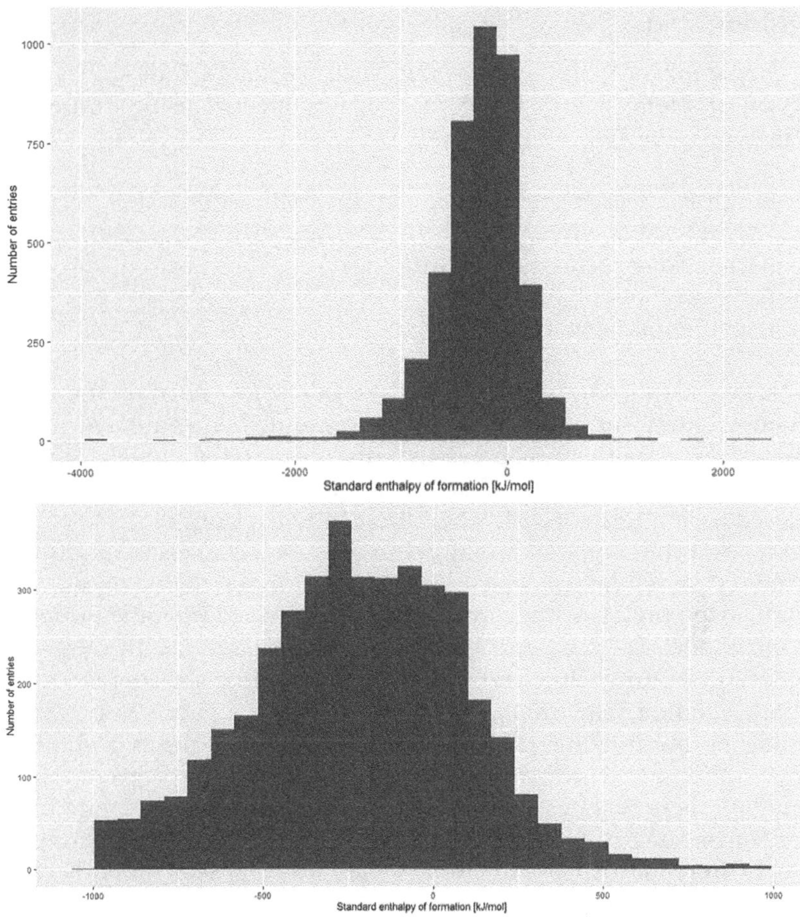

Fig. 3 The distribution of the entries in the data set according to the value of the standard EOF. The upper image shows the individual outliers mentioned in the text outside the value range from $-1000\ \text{kJ}\ \text{mol}^{-1}$ to $1000\ \text{kJ}\ \text{mol}^{-1}$. The lower image shows the distribution within the selected value range.

Table 1 Overview of the descriptors of the training data set.

Feature group	Number of variables	Data type
Aromaticity	1	Logical
Number of atoms	1	Integer
Atom count per element	4	Integer
O_2 balance	1	Float
Number of bonds per bond type	14	Integer
Fingerprints	881	Logical

molecule in a binary format. Each bit in this code represents the presence or absence of certain chemical features in the molecule. In this case, the PubChem format was chosen, which consists of 881 characters.[24] See Table 1 below for an overview of the descriptors for the model training.

3 Modelling

The prediction of a numerical value such as the EOF describes a regression problem. Regression models can be created using a variety of ML methods, the following methods were used:

1. Multivariate linear regression with interactions (MLR)
2. Partial least squares (PLS)
3. Kernel ridge regression[25] (KRR)
4. Random forest[26] (RF)
5. Deep neural network[27] (DNN)

A train-test-split was used to train the models, *i.e.* the models are only trained on a subset of the entire data set and then evaluated on the subset of test data that was previously unknown to the model. The ratio of training to test data was changed several times, the best result was achieved with a ratio of 4 : 1, whereby 3436 entries of the data set were used for training and 859 remained in the test data set. In addition, a 10-fold cross-validation was carried out during the training. The training itself took place according to the supervised learning principle, *i.e.* the model is given the expected target value, *i.e.* the expected standard EOF, for each entry from the training data set. Depending on the ML method, the model then optimizes its parameters during training by minimizing a loss function so that the desired output is achieved. For a DNN this is done by the so-called backpropagation, an input data set is first passed through the network during training in the so-called forward pass, resulting in an initial prediction of the model. An error function or loss function is then used to determine the deviation from the expected value. In the subsequent backward pass, the determined error is fed back from the output of the network, whereby the gradient of the error is determined with respect to each weight in the network (gradient descent). Finally, the weights are updated by subtracting the respective gradient multiplied by the learning rate. The learning rate is dynamically adjusted several times during the training process, when the loss value is stagnating, with the help of a callback function. The mean absolute error (MAE) is used to compare the models and evaluate the accuracy of the predictions.

Table 2 shows the current scoring of the various models. As already mentioned in the previous chapter, the model development took place in several cycles of data preprocessing, model adaptation and model

Table 2 ML-Model scoring comparison.

ML-method	MAE [kJ mol^{-1}]
MLR	77.3
PLS	140.7
KRR	97.3
RF	69.4
DNN	47.1

training, with changes being made in each case to improve the result. In all runs, the RF and DNN models performed best.

3.1 Random forest

The RF method belongs to the ensemble methods of statistics, which are characterized by the aggregation of several individual algorithms. In the case of RFs, this is a composition of several decision trees. When training a RF model, the training data set is divided into several subsets ("bagging") and a decision tree is trained on each subset. This process is repeated several times, and the subsets are shuffled again in each one of them. The result of the entire model is produced at the end according to the majority principle, the average value in the case of a regression, of the predictions of all decision trees. Random forest models are easily scalable and can be applied to both small and large data sets, as only the number and size of the decision trees need to be adjusted. In theory, they are also more robust against outliers than, for example, a linear model, as the effects of incorrect predictions of individual outliers can be minimized by aggregating the results of several trees. This last aspect might also be the reason why the RF-Model performs comparatively well on this data set, since there are still some outliers, which caused more problems for the adjustment of the other models.

3.2 Deep neural network

In its current state, the DNN model is a feed-forward model consisting of three densely connected layers, with a leaky rectified linear unit (ReLU) as activation function. Feed-forward models only allow to transfer information from the input of the network towards the output. Dense layers or also known as fully connected layers, are layers in which every neuron is connected with every neuron of the following layer. In between, there are filter layers, such as Dropout layer and Kernel Regularizer, which are used to prevent an early overfitting during the training process.

3.3 Hyperparameter tuning

An attempt was made to optimize the results of these two best performing models by hyperparameter tuning. In hyperparameter tuning, a parameter space is set up in which a range of values is defined for predefined model settings. Various combinations of parameters from this space are then compiled according to a random function, a grid search or a Bayesian optimization and a model is trained for each combination of that selection. In the case of an RF model, different split rules of the individual trees, the maximum number of trees in the forest and the number of variables required to split, were dynamically adjusted.

A DNN usually has much more possible hyperparameters than a RF depending on the size of the model. Table 3 shows the hyperparameter space for the last iteration of the hyperparameter training of the DNN

Table 3 DNN-Model hyperparameter tuning value range.

Parameter	Value range	Number of options
Dense 1 neurons	8, 16, . . . , 2048, 4096	10
Dense 2 neurons	8, 16, . . . , 2048, 4096	10
Dense 3 neurons	8, 16, . . . , 2048, 4096	10
L2 kernel regularizer weighting factor	10-2, 10-3, 10-4	3
Dropout factor	0.1, . . . , 0.5	5
Learning rate of the optimizer function	10-2, 10-3, 10-4, 10-5	4
Total number of combinations	60 000	

Table 4 Layer sizes and number of trainable parameters of the DNN model.

Layer	Size	Trainable parameters
Input	902	0
Dense 1	4096	3 694 592
Dense 2	2048	8 390 656
Dense 3	512	1 049 088
Output	1	513
Total	7559	13 134 849

model. If one would try to run through all combinations, 60 000 individual models would have to be trained. The training runs on a powerful HPC (Higher Performance Cluster) and yet the time required for 60 000 model trainings would take several weeks, even though the model is still relatively small. For this reason, only a selection of 10% of all possible combinations were chosen and tried out. The values of the individual parameters that occur most frequently are taken from the 100 runs with the lowest MAE on the test data set. The hyperparameter values of the best runs result in a structure as shown in Table 4, with a total of just over 13 million trainable parameters.

The residual plot in Fig. 4 shows all predictions of the DNN model for the entries of the test data set against the respective expected value of the enthalpy of formation. For comparison, the corresponding plots for the other methods can be found in the appendix (Fig. 5–8). With the presented current model architecture and data set, the model achieves a mean absolute error of approx. 47.1 kJ mol^{-1} on the test data set. Overall, the result is already very good for the most part, still some outlies greatly divert from the expected result.

4 Summary and next development steps

This work gave an insight into the efforts of developing and improving machine learning models for the prediction of the standard enthalpy of formation for the development of new energetic materials. Although the existing models have already been significantly improved step by step, by expanding the descriptor set there is still some potential for improvement. With a deep neural network model showing the best

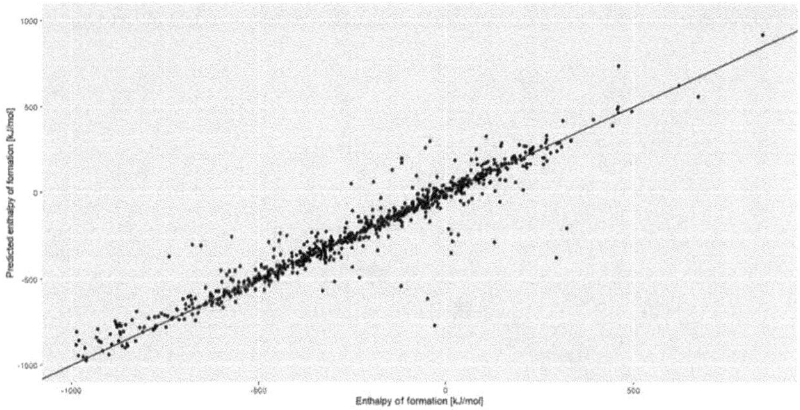

Fig. 4 Residual plot for the model scoring on the test data set with a MAE of 47.1 kJ mol^{-1}.

performance we are currently working on a comparison of this model's predictions with the results of quantum mechanical simulations of the same molecules, to better understand weaknesses of the current design. Some of those issues are already known. The DNN model at its current state has problems with molecules that contain multiple aromatic bonds. This is due to the fact, that the descriptor set only describes the occurrence or absence of aromatic bonds and not a clear number for it. To address this issue, we plan to switch from CDK Objects to another library like the open source RDKit[28] toolkit to get a more precise description of the bond types with regards to ionic and aromatic bonds which should significantly improve the model. The current model also lacks an option for differentiating between different isomers.

Another problem comes with some of the entries of the data set itself. Due to the automated process of uplifting the data set with additional information, there are still false entries in the data set, which lead to false adjustments of the models. These entries are hard to detect without analysing all of the data set's entries manually. Some of these false entries have been identified after reviewing the results of a model on the test set, by closer inspecting the cases with the highest difference to the expected EOF value.

On the modelling part, there are also plans to try out other architecture types. New architectures such as physics-informed neural network (PINNs) or transformer models could help improving the results. The latter became famous through ChatGPT and could be quickly adapted by fine tuning existing models. The first pre-trained models which could be suitable for this purpose already exist, such as MolFormer.[29] PINNs are a promising way of combining the classical "data-driven" ML development approach with a "physics-based" approach,[30] where equations-of-state and partial differential equations can be added to the model's structure or used during the training process.

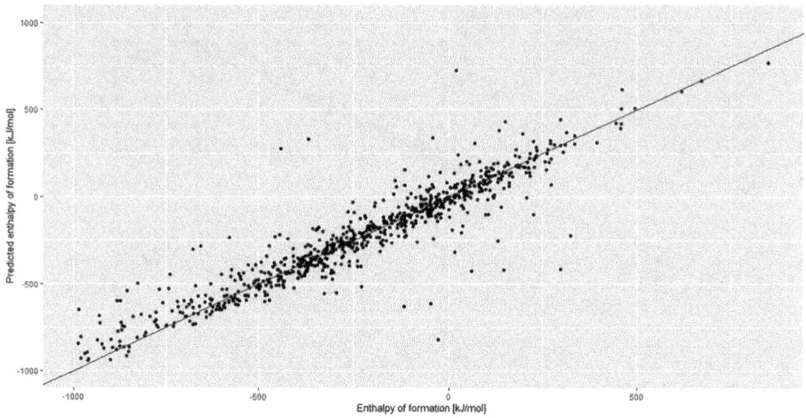

Fig. 5 Residual plot for the MLR model with a MAE of 77.3 kJ mol^{-1}.

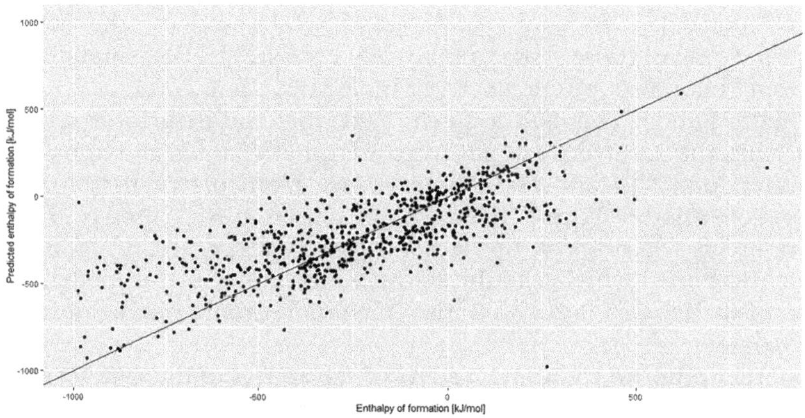

Fig. 6 Residual plot for the PLS model with a MAE of 140.7 kJ mol^{-1}.

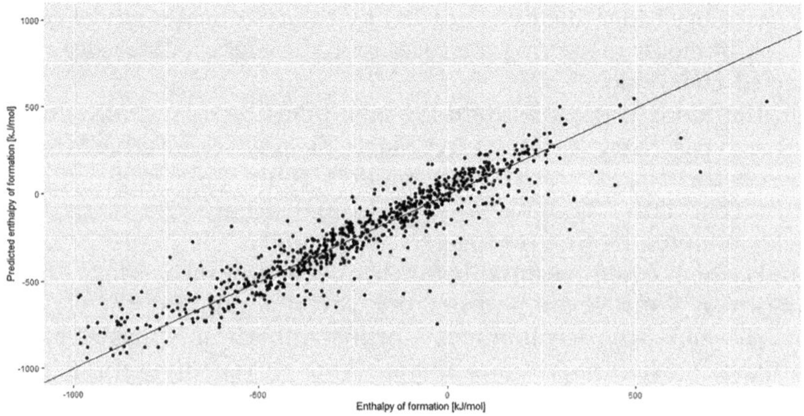

Fig. 7 Residual plot for the KRR model scoring with a MAE of 97.3 kJ mol^{-1}.

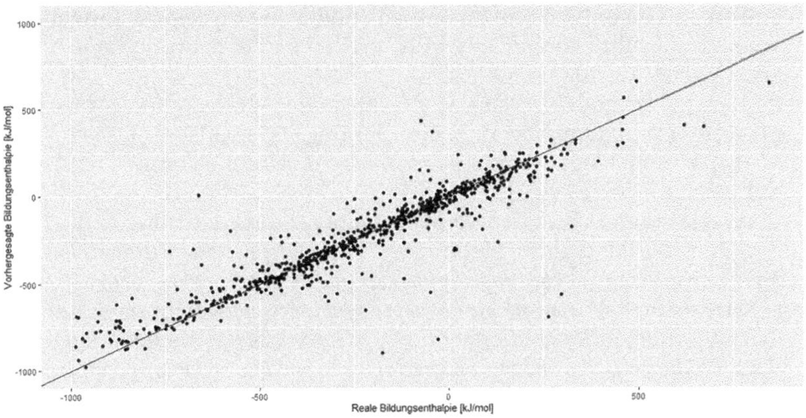

Fig. 8 Residual plot for the RF model scoring with a MAE of 69.4 kJ mol^{-1}.

References

1 E. F. C. Byrd and B. M. Rice, Improved prediction of heats of formation of energetic materials using quantum mechanical calculations, *J. Phys. Chem. A*, 2006, **110**, 1005–1013.

2 B. M. Rice, S. V. Pai and J. Hare, Predicting heats of formation of energetic materials using quantum mechanical calculations, *Combust. Flame*, 1999, **118**, 445–458.

3 J. Kirkpatrick, B. McMorrow and D. H. P. Turban, *et al.*, Pushing the frontiers of density functionals by solving the fractional electron problem, *Science*, 2021, **374**, 1385–1389.

4 C. Chen, Y. Zuo and W. Ye, *et al.*, A Critical Review of Machine Learning of Energy Materials, *Adv. Energy Mater.*, 2020, **10**, 1903242.

5 J. Choi, P. Nguyen and O. Sen, *et al.*, Artificial intelligence approaches for energetic materials by design: state of the art, challenges, and future directions, *Propellants, Explos., Pyrotech.*, 2023, 00276.

6 B. Huang and O. A. von Lilienfeld, Ab Initio Machine Learning in Chemical Compound Space, *Chem. Rev.*, 2021, **121**, 10001–10036.

7 X. Tian, S. Song and F. Chen, *et al.*, Machine learning-guided property prediction of energetic materials: Recent advances, challenges, and perspectives, *Energ. Mater. Front.*, 2022, **3**, 177–186.

8 F. Hou, Y. Ma and Z. Hu, *et al.*, Machine Learning Enabled Quickly Predicting of Detonation Properties of N-Containing Molecules for Discovering New Energetic Materials, *Adv. Theory Simul.*, 2021, **4**, 2100057.

9 B. C. Barnes, D. C. Elton, Z. Boukouvalas, *et al.*, Machine Learning of Energetic Material Properties, in *Proceedings of the 16th International Detonation Symposium*, 2018.

10 H. Wang, Y. Ji and Y. Li, Simulation and design of energy materials accelerated by machine learning, *Wiley Interdiscip. Rev.: Comput. Mol. Sci.*, 2020, **10**, e1421.

11 D. C. Elton, Z. Boukouvalas and M. S. Butrico, *et al.*, Applying machine learning techniques to predict the properties of energetic materials, *Sci. Rep.*, 2018, **8**, 9059.

12 D. Mathieu, Molecular energies derived from deep learning: application to the prediction of formation enthalpies up to high energy compounds, *Mol. Inf.*, 2021, 2100064.

13 F. Wang, J. Liu and D. Tian, *et al.*, Prediction of the enthalpy of formation for aromatic polynitro compounds with artificial neural network, *Chin. J. Energ. Mater.*, 2004, **12**, 207.
14 X. Tian, X. Qi and Y. Wang, *et al.*, Predicting Enthalpy of Formation of Energetic Compounds by Machine Learning: Comparison of Featurization Methods and Algorithms, *Propellants, Explos., Pyrotech.*, 2023, **48**, e202200236.
15 C. Wespiser and D. Mathieu, Application of Machine Learning to the Design of Energetic Materials: Preliminary Experience and Comparison with Alternative Techniques, *Propellants, Explos., Pyrotech.*, 2023, e202200264.
16 S. Kelzenberg, P.-B. Kempa, S. Wurster, *et al.*, New version of the ICT-thermodynamic code, in *Proceedings of the 45th International Annual Conference of the Fraunhofer ICT: Energetic Materials – Particles, Processing, Applications*, ed. Fraunhofer ICT, Karlsruhe, 2014.
17 D. Weininger, SMILES, a chemical language and information system. 1. Introduction to methodology and encoding rules, *J. Chem. Inf. Comput. Sci.*, 1988, **28**, 31–36.
18 PubChem, https://pubchem.ncbi.nlm.nih.gov/ (2024, accessed 25 July 2024).
19 R. T. Fielding, Architectural Styles and the Design of Network-based Software Architectures, PhD thesis, University of California, Irvine, 2000.
20 DeepL Translate: The world's most accurate translator, https://www.deepl.com/en/translator (2024, accessed 25 July 2024).
21 I. Weber, H. Linka, D. Mertens, *et al.*, FhGenie: A Custom, Confidentiality-preserving Chat AI for Corporate and Scientific Use, *arXiv*, 2024, arXiv:2403.00039v1.
22 C. Steinbeck, Y. Han and S. Kuhn, *et al.*, The Chemistry Development Kit (CDK): an open-source Java library for Chemo- and Bioinformatics, *J. Chem. Inf. Comput. Sci.*, 2003, **43**, 493–500.
23 M. Heil and J. Langer, Energetic materials and machine learning, Requirements and application, in *Proceedings of the 52nd International Annual Conference of the Fraunhofer ICT: Energetic Materials – Analysis, Characterization, Modelling*, ed. Fraunhofer ICT, Karlsruhe, 2023, pp. 21-1–21-11.
24 PubChem National Center for Biotechnology Information, PubChem Substructure Fingerprint, https://ftp.ncbi.nlm.nih.gov/pubchem/specifications/pubchem_fingerprints.pdf (accessed 25 July 2024).
25 T. Hofmann, B. Schölkopf and A. J. Smola, Kernel methods in machine learning, *Annu. Stat.*, 2008, **36**, 1171–1220.
26 L. Breiman, Random Forests, *Mach. Learn.*, 2001, **45**, 5–32.
27 J. Schmidhuber, Deep learning in neural networks: an overview, *Neural Networks*, 2015, **61**, 85–117.
28 N. Brown, *In Silico Medicinal Chemistry, Computational Methods to Support Drug Design*, Cambridge, The Royal Society of Chemistry, 2016.
29 J. Ross, B. Belgodere, V. Chenthamarakshan, *et al.*, Large-Scale Chemical Language Representations Capture Molecular Structure and Properties, *arXiv*, 2021, arXiv:2106.09553v3.
30 J. A. Keith, V. Vassilev-Galindo and B. Cheng, *et al.*, Combining Machine Learning and Computational Chemistry for Predictive Insights into Chemical Systems, *Chem. Rev.*, 2021, **121**, 9816–9872.

UCM/MMP cookoff models for explosives containing RDX

Michael L. Hobbs,* Michael J. Kaneshige and William W. Erikson

Determining the thermal response of energetic materials containing RDX (hexahydro-1,3,5-tri-nitro-1,3,5-triazine) is difficult due to pressure dependent reactions that occur within the interior of explosives along with the various binders that decompose simultaneously with the RDX. Recently, we have developed a Universal Cookoff Model (UCM) coupled to a MicroMechanics Pressurization (MMP) model to account for pressure-dependent reactions occurring on the interior of explosives. We assume that the decomposition gases accumulate in uniformly distributed defects that grow spherically until local damage causes the closed pressurizing pore to become open. As the explosive is damaged, the gases can move to other open pores or even into the surrounding head-space. One drawback of the combined UCM/MMP model is insufficient parameters for our RDX based explosives. Here, we use historic data from the Sandia Instrumented Thermal Ignition (SITI) experiment to create UCM/MMP models for seven RDX containing explosives. The models are validated with data from other laboratories.

1 Introduction

Cookoff occurs when an energetic material thermally ignites after being exposed to high temperatures. Exothermic reactions produce energy and gas that cause the temperature and pressure to increase. High temperatures and pressures cause the decomposition rates to accelerate leading to thermal runaway as generated energy is produced faster than it can be dissipated. Reaction acceleration is even more pronounced in materials at high densities where decomposition products accumulate within closed pores causing internal pressurization that is difficult to measure. We have developed a universal cookoff model[1] (UCM) coupled to a micromechanics pressurization (MMP) model[2] to address closed-pore decomposition leading to thermal ignition.

We present UCM/MMP parameters for the following seven RDX explosives: (1) **Comp-B3** (60 wt% RDX and 40 wt% 2,4,6-trinitrotoluene or TNT), (2) **Comp-C4** (91 wt% RDX, 5.3 wt% dioctyl adipate or DOA, 2.1 wt% polyisobutylene, and 1.6 wt% oil, (3) **PBX 9407** (94 wt% RDX and 6% vinyl chloride/chlorotrifluoro ethylene copolymer or VCTFE, (4) **PBXN-109** (64 wt% RDX 20% aluminum, 8% hydroxyl-terminated polybutadiene or HTPB, and 8% DOA), (5) **RDX** (assumed pure), (6) **Semtex 1H** (43 wt% RDX, 43 wt% PETN, 2.5 wt% polyethylene and 11.6 wt% dioctyl sebacate), and (7) **XTX-8004** (80 wt% RDX and 20 wt% Sylgard™).

Sandia National Laboratories, USA. E-mail: mlhobbs@sandia.gov

2 UCM/MMP model

2.1 Energy equation

Predicting thermal ignition requires solving the conductive energy equation,

$$\rho C_{\mathrm{p}} \frac{\partial T}{\partial t} = \nabla \cdot (k \nabla T) + r_{\mathrm{x}} h_{\mathrm{r}} + \dot{q}_{\mathrm{melt}} \tag{1}$$

by including volumetric sources for decomposition ($r_x h_r$) and latent energy sinks $\dot{q}_{melt}$. The momentum equation can also be solved for flow if the explosive melts (*e.g.*, Comp-B3, TNT, PETN, *etc.*). From previous work,[3] we have found that flow effects are hindered by non-melting constituents such as RDX in Comp-B3 that cause the material to stay largely in place. For such situations, flow effects can adequately be accounted for by increasing the thermal conductivity (k) at the melting point. Specific heats (C_p), thermal conductivities (k), and rate coefficients (n, m, E, σ, and λ_m) in eqn (1) can be determined from experiments coupled with finite element simulations.[4] Parameters for RDX containing explosives are given in Tables 1 and 2.

Initially, the universal cookoff model[1] was developed to provide a simple framework for describing thermal ignition of energetic materials. This framework consisted of a single desorption reaction for moisture, two explosive reactions with one being gas-phase dominated and the other being condensed-phase dominated, and a binder reaction. The success of this model is attributed to the flexible form of the distributed rate expressions.

Table 1 Thermophysical properties.[a]

Name	T, K	C_p, J kg^{-1} K^{-1}	T, K	k, W m^{-1} K^{-1} (ρ,[b] kg m^{-3})
Comp-B3	300	1084	300	0.2 (1710)
	350	1240	440	0.2 (1710)
	477	1680	477	0.8[b] (1710)
Comp-C4	300	1004	300	0.16 (1500)
	600	1818	600	0.16 (1500)
PBX 9407	273	1008	400	0.20 (1625)
	482	1681	478	0.20 (1625)
PBXN-109	300	1500	300	0.4 (1650)
	600	1500	600	0.4 (1650)
RDX	300	1004	300	0.20 (1650)
	623	1760	600	0.20 (1650)
Semtex 1H	300	1145	300	0.18 (1530)
	400	1470	500	0.24 (1530)
XTX-8004	298	1100	300	0.2 (1550)
	460	1520	600	0.2 (1550)

[a] Temperature dependent properties are interpolated linearly with constant extrapolation.
[b] Thermal conductivity (k) for Comp-B3 at 477 K was increased artificially from 0.3 to 0.8 W m^{-1} K^{-1} to account for flow. If a true flow model is used, then k at 477 K is 0.3 W m^{-1} K^{-1}.

Table 2 Kinetic parameters.

Name	n	m	E, K	σ, K	RDX: λ_m	PETN: λ_m
Comp-B3	0.35	1.5	24 500	−500	2	—
Comp-C4	0.16	2	25 063	−1000	2	—
PBX 9407	0.23	0.5	22 560	−400	10	—
PBXN-109	0.295	−1	18 012	−400	2	—
RDX	0.295	3	29 772	−400	4	—
Semtex 1H	0[a]	−0.5	17 200	−500	1	4[b]
XTX-8004	0.15	3	28 020	−1000	1	—

[a] No pressure measurements were available to determine n.
[b] Only PETN melt acceleration ($\lambda_m = 4$) was used in the Semtex 1H model.

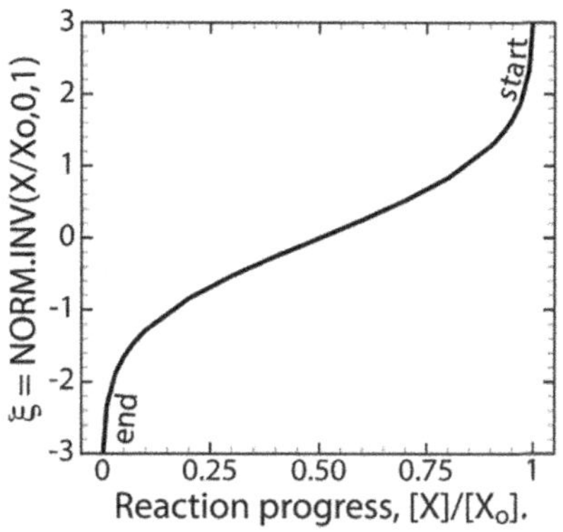

Fig. 1 Inverse of the standard normal distribution.

Two drawbacks of the UCM framework are (1) the difficulty and variability of measuring small amounts of adsorbed moisture and (2) the separation of the explosive reactions from the binder reactions when decomposition products interact. In the current work, we eliminate the moisture desorption step and combined the two explosive reactions with the binder reaction to simplify the UCM framework. This simplified UCM framework is composed of a single first order reactions with respect to the explosive concentration [X]. The rate of explosive decomposition (r_x) is given by eqn (2) with parameters provided in Table 2.

$$r_x = 1.586 \times 10^{15} \left(\frac{P}{P_o}\right)^n \lambda T^m \exp\left[-\frac{(E + \xi\sigma)}{T}\right][\mathrm{X}] \tag{2}$$

2.2 Distributed activation energy

In eqn (2), the inverse of the standard normal distribution (ξ) is used to make the reactions self-accelerating or autocatalytic. The positive sign following the normalized activation energy (E in eqn (2)) causes the activation energy to decrease with the extent of reaction resulting in rate acceleration (for a negative value of σ). Fig. 1 shows the inverse of the standard normal distribution calculated with the Microsoft Excel function "$x = $NORM.INV($X/X_o$,0,1)".

Table 3 Mass fractions of the melting components.

Name	Mush melt: 471–477 K h_l: 160 000 J kg^{-1} ω_{RDX}	350–360 K 98 450 J kg^{-1} ω_{TNT}	413–415 K 177 000 J kg^{-1} ω_{PETN}
Comp-B3	0.60	0.40	0
Comp-C4	0.91	0	0
PBX 9407	0.94	0	0
PBXN-109	0.64	0	0
RDX	1.00	0	0
Semtex 1H	0.43	0	0.43
XTX-8004	0.80	0	0

2.3 Phase change

Most materials that decompose do not melt at a clearly defined temperature but melt over a temperature range referred to as the melty mush zone which begins at the solidus temperature (T_s) and ends at the liquidus temperature (T_L). The model includes RDX melting as well as TNT melting in Comp-B3 and PETN melting in Semtex 1H. Latent enthalpy is accounted for by increasing the specific heat within the phase change temperature range using an effective capacitance method.

The effective capacitance method is shown in eqn (3) where the latent energy is released in a melty mush zone between the solidus temperature (T_s) and the liquidus temperature (T_L) using a Gaussian distribution. In this method, the specific heat is increased artificially within the mush zone to account for both caloric and latent contributions to the transient heat conduction term in the energy equation.

$$C_{\text{eff}} = C_p + \frac{\omega h_l}{\sigma\sqrt{2\pi}} e^{-\frac{z^2}{2}} \quad \text{with} \quad \sigma = \frac{(T_m - T_S)}{3},$$

$$T_m = \frac{(T_L + T_S)}{2}, \quad \text{and} \quad z = \frac{(T - T_m)}{\sigma} \tag{3}$$

The effective capacitance method is used for constituents that melt in each of the explosive formulations. In eqn (3), ω represents the mass fraction of the component that is melting, *e.g.*, ω_{RDX}, ω_{TNT}, or ω_{PETN}, as shown in Table 3. The solidus and liquidus temperature pairs that define the mush melt zone for RDX, TNT, and PETN as well as the latent melting enthalpies (h_l) for pure RDX, TNT, and PETN are also given in Table 3. Other latent effects such as dissolution of RDX in molten TNT[5] or the α-PETN to β-PETN polymorphic phase change[6] are small when compared to melting and are neglected in the current work.

2.4 Phase change induced rate acceleration

Rate acceleration caused by melting is usually significant when the boundary temperature exceeds the melting point of RDX or any other melting component. This acceleration is modelled using a rate multiplier, λ, which is 1 at the solidus temperature (T_s) and λ_m at the liquidus temperature (T_L). The multiplier, λ_m, is given in Table 2. The cosine ramp

Table 4 Explosives containing RDX.

Name	Comp. (wt%), formula	Mw_x, $g\,mol^{-1}$	O_2 bal.	ρ_{co}, $kg\,m^{-3}$	h_f, $kJ\,mol^{-1}$	D, $km\,s^{-1}$	T_a, K
Comp-B3	RDX/TNT (60/40) $C_{4.58}H_{5.6}N_{4.82}O_6$	224.2	−42.6	1742	16.9	7.81	2651
Comp-C4	RDX/DOA/PIB/Oil 91/5.3/2.1/1.6 $C_{3.83}H_{7.5}N_{5.2}O_{5.23}$	211.5	−46.1	1659	16.8	7.90	2471
PBX 9407	RDX/VCTFE (94/6) $C_{3.1}H_{5.9}N_{5.6}O_{5.6}Cl_{0.18}F_{0.18}$	221.8	−24.2	1799	18.4	8.37	2880
PBXN-109	RDX/AL/HTPB/DOA 64/20/8/8 $C_{1.61}H_{2.94}N_{1.44}O_{1.51}AL_{0.62}$	83.4	−78.7	1656	−4.7	7.24	2033
RDX	RDX (100) $C_3H_6N_6O_6$	222.1	−21.6	1806	69	8.78	2930
Semtex 1H	RDX/PETN/ Binder/Plastici. 65.5(max)/ 25(min)/2.5/11.6 or 43/43/2.5/11.5 $C_{4.8}H_{8.86}N_{3.8}O_{6.51}$	224.0	−53.7	1581	−226	7.45	2164
XTX	RDX/Sylgard™ 182 (80/20) $C_{2.57}H_6N_{3.44}O_{3.86}Si_{0.43}$	158.9	−51.8	1579	−176	7.72	2260

function in eqn (4) was used to transition between these two temperatures for RDX:

$$\lambda = \begin{cases} 1, & T \leq T_S \\ 1 + 0.5\left(1 - [\lambda_m - 1]\cos\left[\pi\frac{(T - T_S)}{(T_S - T_L)}\right]\right) & T_S < T < T_L \\ \lambda_m & T \geq T_L \end{cases} \tag{4}$$

2.5 Explosive decomposition

The initial explosive compositions are given in Table 4. Other properties in this table show differences in the various RDX formulations, including the molecular formula, the explosive's molecular weight (Mw_x), the oxygen balance (O_2 bal.) assuming C forms CO_2 and H forms H_2O, the theoretical maximum density (TMD, ρ_{co}), the heat of formation (h_f), the detonation velocity (D) determined at TMD, and the low-pressure adiabatic flame temperature (T_a). The detonation velocities and adiabatic flame temperatures were predicted with the JCZ3 equation of state (EOS) using the TIGER code with the JCZS3 thermodynamic database.[7]

TIGER was also used to predict the equilibrium product hierarchy shown in Table 5 for each of the explosives at 400 K and 1 bar. This thermodynamic state is representative of typical cookoff conditions and doesn't change much, even at elevated pressures. A simplified reaction scheme is also given in Table 4 showing formation of a single gas and various condensed species:

$$X \rightarrow \alpha G + \beta Al + \gamma C + \delta SiO_2. \tag{5}$$

Table 5 Reaction hierarchy for pre-ignition reactions.

Name	Equilibrium reactions, heats for formation (h_f), and reaction enthalpies (h_{rxn})
Comp-B3	$C_{4.58}H_{5.6}N_{4.82}O_6 \rightarrow 2.54\ H_2O + 2.41\ N_2 + 1.73\ CO_2 + 0.13\ CH_4 + 2.72\ C$ Comp-B3→6.81 G + 2.72 C $Mw_G = 28.1\ g\,mol^{-1}$; $h_{f,G} = -191.5\ kJ\,mol^{-1}$ $h_{f,Comp\text{-}B3} = 16.9\ kJ\,mol^{-1}$; $h_r = -1321\ kJ\,mol^{-1}$ (exothermic)
Comp-C4	$C_{3.83}H_{7.5}N_{5.2}O_{5.23} \rightarrow 3.14\ H_2O + 2.60\ N_2 + 1.09\ CO_2 + 0.3\ CH_4 + 0.01\ H_2 + 2.43\ C$ Comp-C4→7.14 G + 2.44 C $Mw_G = 25.5\ g\,mol^{-1}$; $h_{f,G} = -170\ kJ\,mol^{-1}$ $h_{f,Comp\text{-}C4} = 16.0\ kJ\,mol^{-1}$; $h_r = -1226\ kJ\,mol^{-1}$ (exothermic)
PBX 9407	$C_{3.1}H_{5.9}N_{5.6}O_{5.6}Cl_{0.18}F_{0.18} \rightarrow 2.51\ H_2O + 2.82\ N_2 + 1.57\ CO_2 + 0.18\ HCL + 0.18\ HF + 0.13\ CH_4 + 1.40\ C$ PBX 9407→7.37 G + 1.40 C $Mw_G = 27.8\ g\,mol^{-1}$; $h_{f,G} = -176\ kJ\,mol^{-1}$ $h_{f,PBX9407} = 18.4\ kJ\,mol^{-1}$; $h_{rxn} = -1318\ kJ\,mol^{-1}$ (exothermic)
PBXN-109	$C_{1.61}H_{2.94}N_{1.44}O_{1.51}AL_{0.62} \rightarrow 1.08\ H_2O + 0.72\ N_2 + 0.215\ CO_2 + 0.195\ CH_4 + 0.62\ AL + 1.2\ C$ (The aluminum is non-reactive in pre-ignition calculations) PBXN-109→2.21 G + 0.62 AL + 1.20 C $Mw_G = 23.6\ g\,mol^{-1}$; $h_{f,G} = -163\ kJ\,mol^{-1}$ $h_{f,PBXN109} = -4.7\ kJ\,mol^{-1}$; $h_r = -355.6\ kJ\,mol^{-1}$ (exothermic)
RDX	$C_3H_6N_6O_6 \rightarrow 2.7\ H_2O + 3\ N_2 + 1.65\ CO_2 + 0.15\ CH_4 + 1.2\ C$ RDX→7.5 G + 1.2 C $Mw_G = 27.7\ g\,mol^{-1}$; $h_{f,G} = -175\ kJ\,mol^{-1}$ $h_{f,RDX} = 69\ kJ\,mol^{-1}$; $h_r = -1382\ kJ\,mol^{-1}$ (exothermic)
Semtex 1H	$C_{4.8}H_{8.86}N_{3.8}O_{6.51} \rightarrow 3.75\ H_2O + 1.9\ N_2 + 1.38\ CO_2 + 0.34\ CH_4 + 3.08\ C$ Semtex 1H→7.37 G + 3.08 C $Mw_G = 25.4\ g\,mol^{-1}$; $h_{f,G} = -200\ kJ\,mol^{-1}$ $h_{f,Semtex\ 1H} = -226\ kJ\,mol^{-1}$; $h_r = -1249\ kJ\,mol^{-1}$ (exothermic)
XTX-8004	$C_{2.57}H_6N_{3.44}O_{3.86}Si_{0.43} \rightarrow 2.0\ H_2O + 1.72\ N_2 + 0.40\ CO_2 + 0.40\ CH_4 + 1.77\ C + 0.43\ SiO_2$ XTX-8004→4.72 G + 1.77 C + 0.43 SiO_2 $Mw_G = 23.7\ g\,mol^{-1}$; $h_{f,G} = -152.4\ kJ\,mol^{-1}$, $h_{f,SiO_4} = -911\ kJ\,mol^{-1}$ $h_{f,XTX\text{-}8004} = -175.5\ kJ\,mol^{-1}$; $h_r = -935.2\ kJ\,mol^{-1}$ (exothermic)

The explosive concentration [X] can be determined with $\frac{dX}{dt} = -r_x$ with $[X]_o = w_x r_{bo}/Mw_x$. In eqn (5), X represents the explosive. G, Al, C, and SiO_2 represent decomposition gases, aluminum, carbon, and silicon dioxide, respectively. The stoichiometric coefficients (α, β, γ, and δ) are given in Table 5. For example, when X = Comp-B3, $\alpha = 6.81$, $\beta = 0$, $\gamma = 2.72$ and $\delta = 0$.

2.6 Micromechanics pressurization

The conductive energy equation (eqn (1)) is coupled to a micromechanics pressurization (MMP) model through pressure (P) in eqn (2). The MMP model is based on decomposition gases accumulating within a defect as

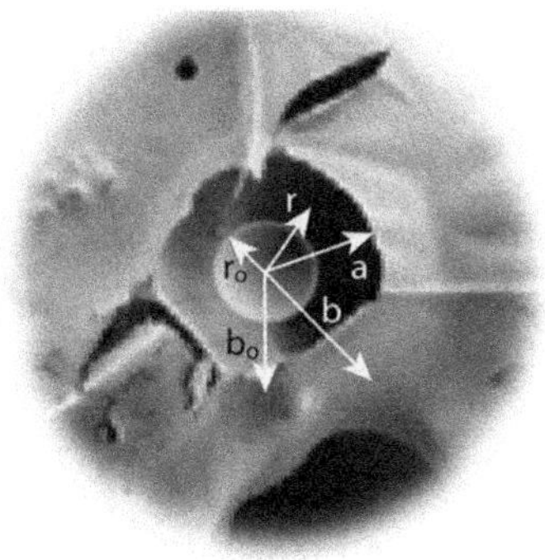

Fig. 2 Pore defect.

Table 6 MMP parameters.

Name	ρ_{co}, kg m^{-3}	S_L, m s^{-1}	S_S, m s^{-1}	K, GPa	E, GPa	n	β, K^{-1}	ϕ_{rdx}	$\phi_{rdx}\Delta\tilde{\varepsilon}_m$, RDX
Comp-B3	1742	3.10	1.68	10.2	12.7	0.29	0.000164	0.579	0.038
Comp-C4	1659	3.26	1.84	10.2	14.2	0.27	0.000270	0.836	0.054
PBX 9407	1799	3.04	1.77	9.10	14.0	0.24	0.000198	0.937	0.061
PBXN-109	1656	0.38	0.053	0.233	0.014	0.49	0.000340	0.587	0.038
RDX	1806	3.52	2.08	12.0	19.3	0.23	0.000190	1.000	0.065
Semtex 1H	1581	3.41	1.98	10.1	15.5	0.24	0.000370	0.376	0.024
XTX-8004	1579	3.41	1.98	3.4	3.2	0.34	0.000300	0.699	0.045

shown in Fig. 2. where the initial defect is shown as a spherical cloud with various arrows describing how the defect changes inside until it ultimately fails.

In Fig. 2, "r_o" represents the initial defect radius, "r" represents the radius of the pore after decomposition but prior to mechanical displacement, and "a" represents the pore radius after decomposition and includes mechanical displacement. The parameters b_o and b are related to the distance between two adjacent nucleation sites. The MMP model uses a simple damage criterion where the pore pressure is used to differentiate between open and closed pores. Gases can only migrate between open pores and the headspace volume where pressure is typically measured. In contrast, gases within closed pores do not co-mingle.

The MMP model provides the local pressure by solving the displacement equations for a symmetric linear elastic material surrounding a pressurizing hollow sphere. The MMP parameters are listed in Table 6 are required to determine the spatially dependent pressure. Derivation of the MMP model equations is beyond the scope of the current work but can be found in ref. 2.

The MMP parameters for the RDX-based explosives include bulk modulus (K), Young's modulus (E), Poisson's ratio (n), distance between nucleation sites ($b_o = 0.000226$ m), pore failure pressure ($P_{fail} = 5 \times 10^6$ Pa), volumetric expansion coefficient (b) used to calculate thermal strain ($\tilde{\varepsilon}_T = \beta(T - T_o)$), and the volume fraction weighted mechanical strain, $\Delta\tilde{\varepsilon}_m$, which is caused by melting. Mechanical strain caused by polymorphic phase changes are assumed to be negligible. K, E, and n were estimated from longitudinal (S_L) and shear (S_S) sound velocities. S_L and S_S were either obtained from ref. 8–11 or estimate using a linear correlation between crystal density (TMD) and sound speed. The volumetric coefficient of thermal expansion (β) was estimated using a linear correlation between K and β using data from the other explosives.[8]

The mechanical strain caused by melting for pure RDX, TNT, and PETN was determined with eqn (6):

$$\Delta\tilde{\varepsilon}_m = \frac{\rho_{co}}{[1+\beta(T_m - 300)]\rho_L} - 1 \tag{6}$$

The mechanical strain is zero below the solidus temperature and $\Delta\tilde{\varepsilon}_m$ above the liquidus temperatures. This volume fraction weighted strain is transitioned using a cosine ramp function like shown in eqn (4). For Comp-B, the additional mechanical strain associated with TNT is 0.048. For Semtex 1H, the additional mechanical strain associated with PETN is 0.031.

Additional MMP parameters include ρ_{co} which represents the crystal density (TMD) at room temperature (300 K) for pure RDX (1806 kg m^{-3}), TNT (1654 kg m^{-3}), and PETN (1778 kg m^{-3}). β represents volumetric expansion coefficient for pure RDX (0.00023 K^{-1}),[12] TNT (0.00016 K^{-1}),[12] and PETN (0.00025 K^{-1}).[12] ρ_L represents liquid densities for RDX (1630 kg m^{-3}),[13] TNT (1472 kg m^{-3}),[14] and PETN (1600 kg m^{-3}, guessed). $\Delta\tilde{\varepsilon}_m$ is the mechanical strain due to melting for pure RDX (0.065), TNT (0.114), and PETN (0.080). The volume fraction of RDX is given in Table 6. The volume fraction of TNT in Comp-B3 is 0.579. The volume fraction of PETN in Semtex 1H is 0.382.

3 SITI calibration and predictions

The Sandia Instrumented Thermal Ignition (SITI) experiment shown in Fig. 3(a) was used to determine the thermal conductivities listed in Table 1 and the reaction parameters listed in Table 2 using measured temperatures at various radial positions. Calibration usually requires both a vented and sealed experiment. The remaining SITI experiments as

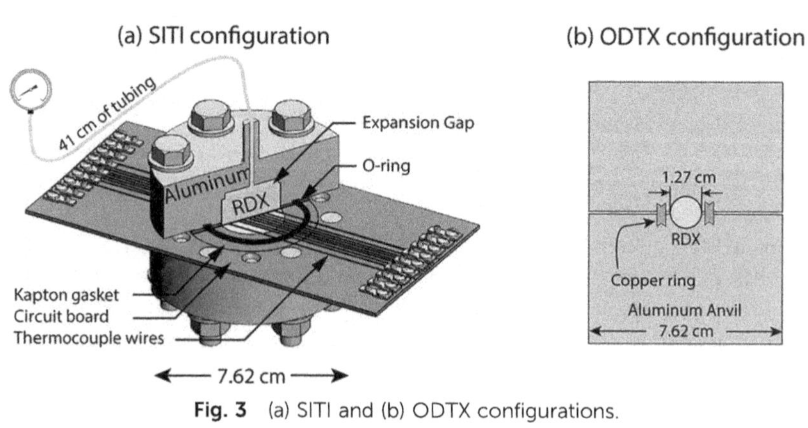

Fig. 3 (a) SITI and (b) ODTX configurations.

well as the One-Dimensional Time-to-eXplosion (ODTX) experiments are used for validation. The ODTX experiment confines a 1.27 cm diameter sphere within constant temperature anvils as shown in Fig. 3(b).

3.1 SITI and ODTX configurations

Typically, the SITI experiment confines two 2.54 cm diameter by 1.27 cm tall cylinders of HMX-based explosive in aluminum with nine type K 127 mm (0.005 in.) diameter thermocouples with measuring points located at radial positions in mm of 0, 1.70, 2.55, 3.40, 4.25, 5.11, 5.96, 8.81, and 11.7 and placed between the two explosive cylinders. The outer surface of the 7.62 cm diameter by 4.58 cm tall aluminum confinement is heated using rope heaters controlled by a thermocouple on the lateral surface. The external aluminum surface is heated to a set point temperature, T_{sp}, in 600 s and held until the explosive thermally ignites. Two expansion

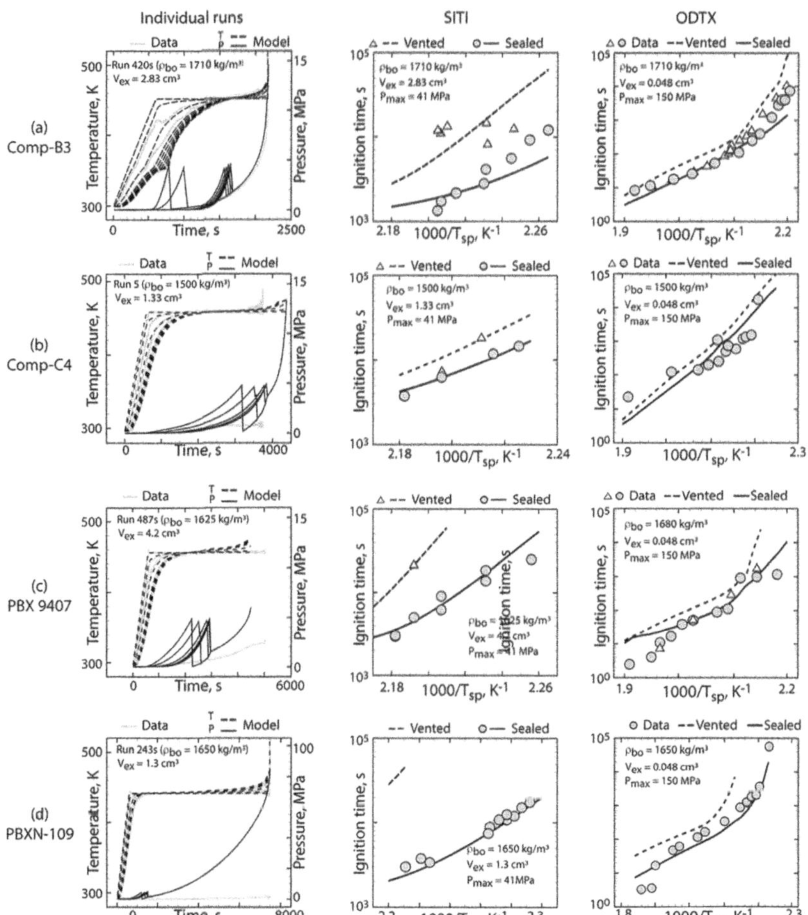

Fig. 4 Individual runs showing predicted and measured pressure, thermal ignition time for the SITI experiments, and thermal ignition times for the ODTX experiments for (a) Comp-B3, (b) Comp-C4, (c) PBX 9407, and (d) PBXN-109. More detail regarding the SITI data can be found in ref. 3 and 16–18. Details regarding the ODTX data can also be found in ref. 1, 3, 19 and 20.

gaps that are above and below the explosive are also machined into the confining aluminum. Each expansion gap has a diameter of 2.22 cm and is 0.16 cm tall. The top expansion gap is connected to a calibrated Kulite Model HEM-375 pressure transducer.

ODTX experiments measure thermal ignition times in a 1.27 cm diameter sphere of explosive.[15] Two preheated 7.62 cm diameter by 5.08 cm high aluminum cylinders hydraulically confine a spherical sample to 150 MPa in a machined cavity with a knife edge channel surrounding the explosive. A copper gasket is used to retain the decomposition gases within the cavity. At time zero, preheated anvils are quickly placed around the explosive providing a constant temperature boundary temperature.

3.2 Model and data comparisons

The model was used to predict both internal temperature/pressure profiles as well as thermal ignition time for seven explosives as shown in Fig. 4 and 5. The plots on the left-hand side of these figures show a single individual SITI run from each explosive. The individual runs provide the measured/predicted temperature profiles at the nine thermocouple locations as well as the predicted pressure profiles at these radial locations. With the outer locations reaching high pressures prior to inner locations. The pressure profiles show a gradual increase followed by an abrupt

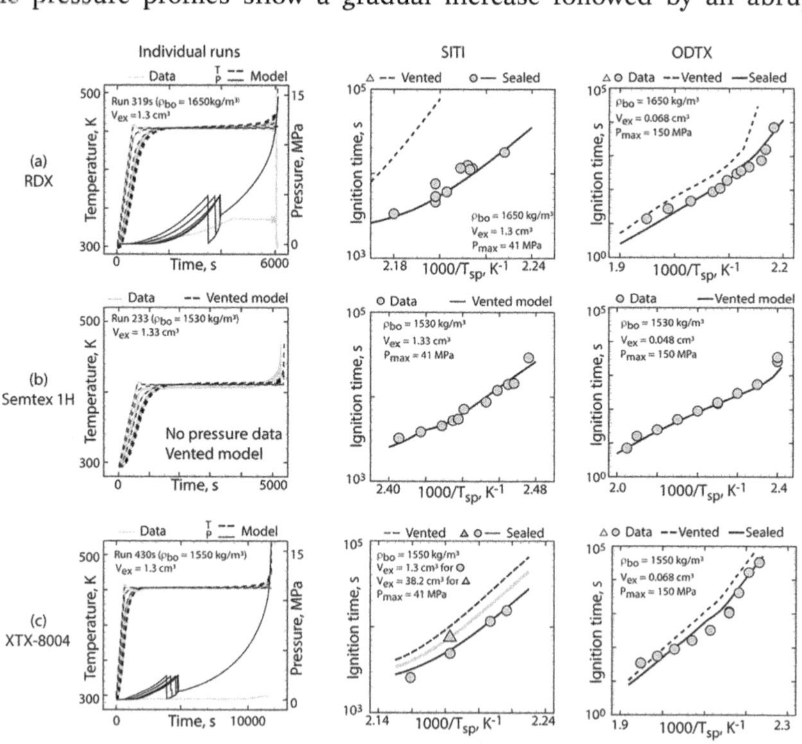

Fig. 5 Individual runs showing predicted and measured pressure, thermal ignition time for the SITI experiments, and thermal ignition times for the ODTX experiments for (a) RDX, (b) Semtex 1H, and (c) XTX-8004. More detail regarding the RDX SITI data can be found in ref. 21. Details regarding the ODTX data can be found in ref. 19, 22 and 23.

decrease in pressure as the material is thermally damaged and transitions from a closed-pore structure to an open pore structure. Pressure in the open pore explosive is the same as in the expansion gap which is measured with the pressure transducer.

Fig. 4 and 5 also provide the thermal ignition time for each of the seven explosives within the SITI configuration (middle plots) and the ODTX configuration (plots on right-hand side of the Fig. 4 and 5). Predicted ignition times for both the SITI experiments and the ODTX experiments show good agreement. A full uncertainty analysis for the RDX and PBXN-109 can be found in ref. 16 indicating that the uncertainty is small in predicted ignition times even when input parameters are adjusted by ±5% of the mean values.

4 Summary and conclusions

The UCM/MMP model has been significantly simplified and used to model thermal ignition of seven explosives containing RDX: (1) Comp-B3, (2) Comp-C4, (3) PBX 9407, (4) PBXN-109, (5) RDX, (6) Semtex 1H, and (7) XTX-8004. The kinetic model consists of a single first order decomposition reaction using a pressure dependent distributed reaction rate based on the reaction extent. Pressure was determined as a field variable using a micromechanics pressurization model. The equilibrium product hierarchy was determined using the TIGER equilibrium code[7] with the product gas molecular weight and heat of formation determined using a mole fraction based mixing rule. The model was calibrated with both vented and sealed SITI experiments and validated with additional SITI experiments as well as ODTX experiments.

The SITI data was taken between 2004–2017: (1) 2013 for Comp-B, (2) 2007 Comp-C4, (3) 2016 for PBX 9407, (4) 2002 for PBXN-109, (5) 2004 for RDX, (6) 2004 for Semtex, and (7) 2015 for XTX-8004. Although we attempted to measure pressure in these experiments, we were not always successful. Primary problems included plugging of the pressure tubing which prevented data acquisition for the sealed tests. Plugging of the vent port may have also caused data interpretation errors for the vented tests. The best measurements were for the Comp-B experiments. In fact, no pressure measurements were obtained for the Semtex 1H experiments and the pressure exponent was set to zero. Pressure measurements were also inadequate for PBXN-109 and the pressure exponent was assumed to be the same as for the RDX model.

Although the model adequately matched all the older cookoff data, there is a need for better small-scale cookoff experiments that are well characterized and include internal temperature as well as pressure. Such experiments should include details regarding any headspace that could fill with decomposition gases. Care should be taken to prevent restriction of the pressure tubing. Experiments should consider investigating cookoff in low density, even shredded, explosives to prevent holdup of gases within the interior of the explosives. Higher density experiments that match the intended applications can be used for model validation and for further safety calculations.

References

1 M. L. Hobbs, M. J. Kaneshige and W. W. Erikson, in *Energetic Materials—Past, Present and Future*, Fraunhofer-Institut für Chemische Technologie ICT, 2019, vol. 9, pp. 1–12.

2 M. L. Hobbs, J. A. Brown, M. J. Kaneshige and C. Aviles-Ramos, *Propellants, Explos., Pyrotech.*, 2022, **47**, e202100156.

3 M. L. Hobbs, M. J. Kaneshige, W. W. Erikson, J. A. Brown, M. U. Anderson, S. N. Todd and D. G. Moore, *Combust. Flame*, 2020, **215**, 36.

4 Z. D. Lawless, M. L. Hobbs and M. J. Kaneshige, *J. Energ. Mater.*, 2020, **38**(2), 214.

5 M. L. Hobbs, M. J. Kaneshige, S. N. Todd and T. R. Krawietz, *J. Therm. Anal. Calorim.*, 2020, **142**, 861.

6 M. L. Hobbs and M. J. Kaneshige, *Combust. Flame*, 2022, **237**, DOI: 10.1016/j.combustflame.2021.111877.

7 M. L. Hobbs, R. G. Schmitt, H. K. Moffat, and Z. Lawless, in *Sixteenth International Detonation Symposium*, Office of Naval Research publication number ONR-43-5762-19, 2018, p. 651.

8 S. P. March, in *LASL Shock Hugoniot Data*, University of California Press, Berkeley, 1985.

9 B. M. Dobratz and P. C. Crawford, in *LLNL Explosives Handbook Properties of Chemical Explosives and Explosive Simulants*, Lawrence Livermore National Laboratory Report UCRL-52997, 1981.

10 J. J. Haycraft, L. L. Stevens and C. J. Eckhardt, *J. Chem. Phys.*, 2006, **124**, 024712.

11 M. A. McClelland, T. D. Tran, B. J. Cunningham, R. K. Weese, and J. L. Maienschein, in *Insensitive Munitionas & Energetic Materials Technology Symposium*, 2003.

12 T. R. Gibbs and A. Popolato, in *LASL Explosive Property Data*, California Press, Berkeley, 1980.

13 D. W. Carlson, G. W. Nauflett, J. W. Brasch and T. D. Austin, in *17th JANNAF Combustion Meeting*, Hampton, 1980.

14 D. M. Dattelbaum, R. S. Chellappa, P. R. Bowden, J. D. Coe and M. A. Margevicius, *Appl. Phys. Lett.*, 2014, **104**, 021911.

15 P. C. Hsu, G. Hust, C. May, M. Howard, S. K. Chidester, H. K. Springer and J. L. Maienschein, *AIP Conf. Proc.*, 2012, **1426**, 559.

16 M. L. Hobbs, M. J. Kaneshige and W. W. Erikson, in *JANNAF 34th ESHS Joint Meeting*, Salt Lake City, 2023.

17 M. L. Hobbs, M. J. Kaneshige, W. W. Erikson, J. A. Brown, M. U. Anderson, S. N. Todd and D. G. Moore, *Combust. Flame*, 2020, **213**, 268.

18 M. L. Hobbs, M. J. Kaneshige and C. D. Yarrington, *Combust. Flame*, 2018, **198**, 278.

19 P. C. Hsu, G. Hust, W. M. Howard and J. L. Maienschein, in *Fourteenth International Detonation Symposium*, Office of Naval Research publication number 351-10-185, 2010, p. 984.

20 P. C. Hsu, S. Strout, M. McClelland and F. Ellsworth, in *Lawrence Livermore National Laboratory Report LLNL-TR-681467*, 2016.

21 M. L. Hobbs, M. Steyskal and M. J. Kaneshige, in *JANNAF 24th Propulsion Systems Hazards Subcommittee Meeting*, Newton, 2008.

22 R. R. McGuire and C. M. Tarver, in *Seventh Symposium (International) on Detonation*, Naval Surface Weapons Center publication number NSWC MP 82-334, 1981, p. 56.

23 P. C. Hsu, S. Strout, A. Nye and R. Ganeriwala, in *Lawrence Livermore National Laboratory Report LLNL-TR-771282*, 2019.

Monte Carlo analysis for hot-spot burning in generalized polygons

Adam Schofield,*[a] William Belfield[a] and John P. Curtis[a,b]

When modelling the burning of high explosives, it is commonly assumed that hot-spots burn outwards with a spherical burn front originating at the centre of a spherical explosive crystal (grain). Under this assumption, once ignited, the spherical burn front progresses outwards, and the entire grain is consumed when the burn front reaches the grain surface. In a complimentary paper, we have derived the analytical solution to a centrally ignited spherical burn front inside a cubic explosive crystal, assuming a constant burn front speed. Taking inspiration from this, here we use a Monte Carlo method and extend the work to consider spherical burn fronts spreading through crystals of arbitrary polygon shapes. A Voronoi tessellation method is used to generate a distribution of polygon shapes which may be used as a representation of the crystals that make up a high explosive. The characteristic burn times of these polygons are compared to the burn time of spherical crystals with equal volume or equal specific surface area. Off-centred hotspots are also investigated, looking at the differences between an edge ignited hotspot (driven by inter-crystalline phenomena such as frictional heating between neighbouring crystals) *versus* the centrally ignited hotspots (driven by intra-crystalline mechanisms, such as localized slip-on crystallographic planes and/or internal cracking).

1 Introduction

Knowledge and understanding of how high explosives burn is of key importance for safety modelling and hazard prediction, but understanding of when an ignition may cause a violent reaction is incomplete. The process of an explosive burning and growing to detonation is known as the Deflagration-to-Detonation Transition (DDT), previous work[1] has shown a considerable dependence for the explosives response to burning on the specific surface area. Many models[2,3] calculating sub-sonic burning after ignition work under the assumption that the time to full burn given a laminar burn speed for an element is derived from that of a similar sphere having the same volume – for this sphere,

$$S_C = A_C/V_C = 3/r_S. \tag{1}$$

where S_C is the surface area to volume ratio of the element (or crystal), such that A_C is the surface area and V_C is the volume, and r_S is the radius of the similar sphere for that element.

It is commonly assumed that a centrally ignited hot-spot burns outwards with a spherical burn front. Under the above assumption of a similar sphere, the elements time to full burn is taken as the time it takes this burn front to reach the similar spherical surface with a constant burn speed. In reality the crystals that make up a high explosive are often

[a] *Materials Physics Group, AWE Nuclear Security Technologies, RG7 4PR, UK. E-mail: adam.schofield@awe.co.uk*
[b] *Department of Mathematics, University College London, Gower Street, London, WC1E 6BT, UK*

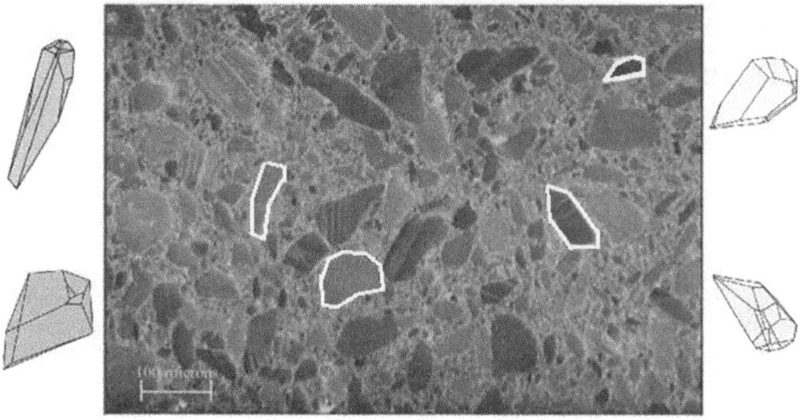

Fig. 1 Micrograph of explosive microstructure showing non spherical crystal shapes and some representative crystal examples.

not spherically shaped and hence may not burn up as their similar spheres do.

Fig. 1 shows a range of shapes present in the microstructure of a high explosive, showing the broad distribution of some highly non spherical shapes. As can be seen from the representative crystal shapes around the edges,[4] the crystals can range from more spherical (bottom left) to increasingly non spherical (top left). The time to full burn consumption is computed over a distribution of polygons and these times are compared with the assumption that the crystals burn proportional to their similar spheres.

2 Methodology

A complementary sister paper[3] derived an analytical solution for the outward burn of a cubic shaped crystal with a centrally ignited spherical burn front and compared these times to that of similar spheres. A Monte Carlo tool was developed to check against this analytical solution. It was found to sufficiently converge to the analytical solution using 10 000 randomly generated Monte Carlo points as demonstrated in Fig. 2.

Although the computational time required to find a solution is greatly increased by the repeat sampling nature of the Monte Carlo solver, it is recognised that the generality of the solver allows it to extend to much more complex geometries and scenarios. Hence a Voronoi tessellation method is used to generate a distribution of polygons which can represent the crystallographic microstructure of a high explosive.

Fig. 3 shows a few of the different geometries that are generated *via* the Voronoi tessellation, each showing different non spherical features that will slow down the time to be fully consumed by a spherical burn front.

The metric used to compare each of these crystals to their similar spheres (and to each other) is the sphericity of the polygon. This is a measure of how closely the shape of an object resembles that of a sphere.

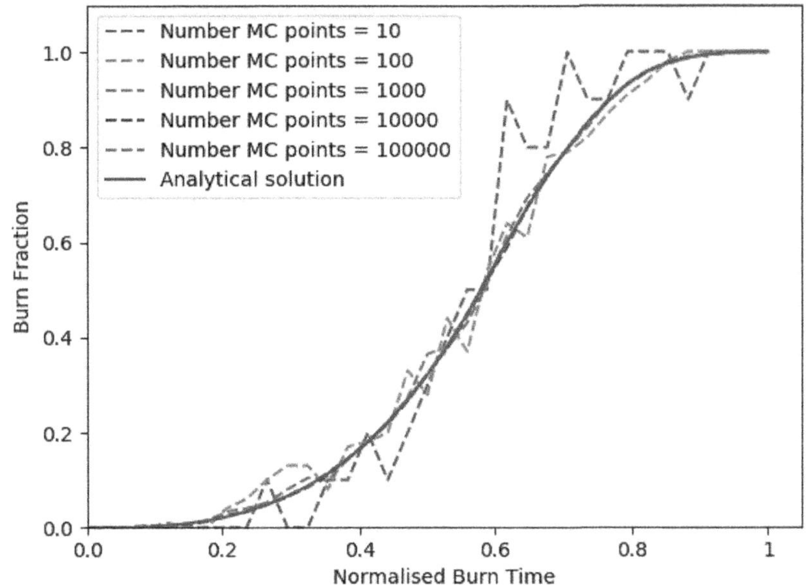

Fig. 2 Comparison and agreement of the Monte Carlo solver with the analytical solution.

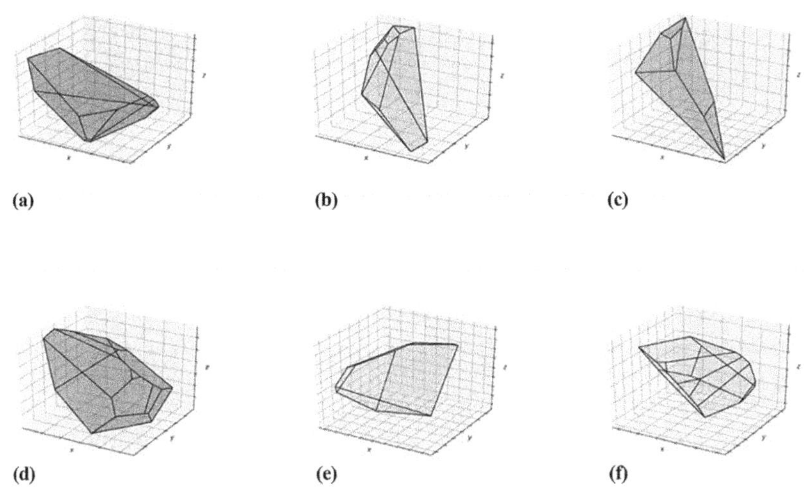

Fig. 3 Example polygonal shapes from a Voronoi tessellation used to represent crystals.

It is the scaled ratio of the surface area and the volume of the polygon, such that a perfect sphere will have a sphericity of unity and less spherically shaped polygons will have decreasing sphericities. The formula is given,

$$\Psi = \left(36\pi {V_C}^2\right)^{1/3} / A_C. \tag{2}$$

3 Centrally ignited hot-spot burning

Centrally ignited hot-spots are driven by intra-crystalline mechanisms, such as localised slip-on crystallographic planes and/or internal cracking.

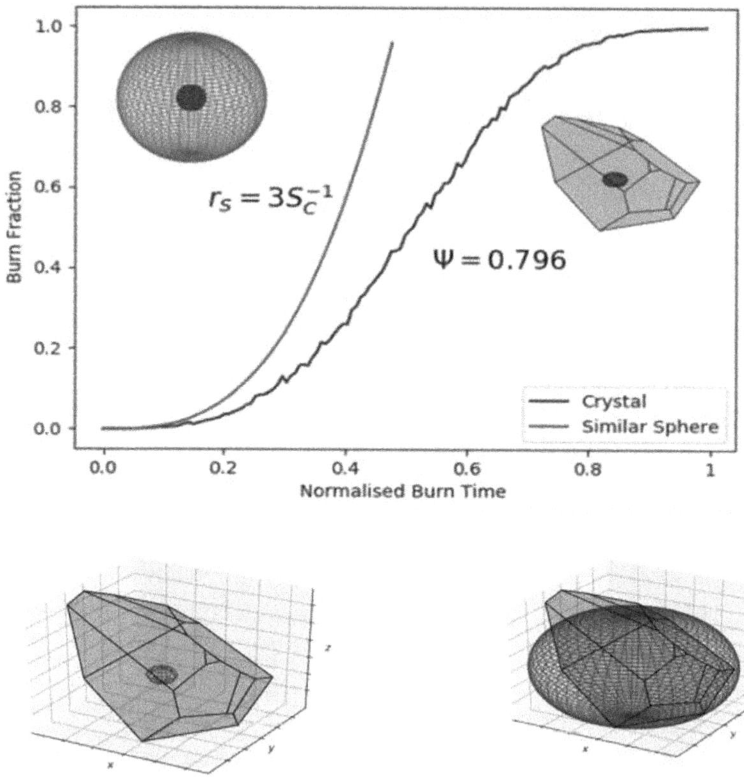

Fig. 4 Crystal with a spherical burn front advancing from the centre with burn times compared to that of the similar sphere.

As all generated crystals are simple polygons, the geometric centre coordinate is found *via* averaging over the polygon's vertex cartesian coordinates. A spherical burn front is then advanced at a constant speed and the burn fraction of the crystal can be calculated over a range of radii using the Monte Carlo solver.

It can be seen from Fig. 4 that even for a crystal of ~0.8 sphericity, the crystal will take ~2.5 times as long to be fully burnt as the similar sphere. That is the ratio of time to full burn for a similar sphere to the crystal, $t_{\mathrm{bS}}/t_{\mathrm{bC}} = 0.4$. Looking at the shape of the burn time curve in Fig. 4, initially the slope is equivalent to that of the burning similar sphere, but at later times slows off dramatically as only as small portion of elongated crystal remains unburnt.

The $t_{\mathrm{bS}}/t_{\mathrm{bC}}$ metric is a useful one as it gives a comparison value for the differing crystal shapes to compare against the sphericities. Looking back at Fig. 3, the burn time ratios and sphericities are tabulated below.

It can be seen from Table 1 that, as expected, crystals that have a lower sphericity reach full burn consumption in much slower times than that of their similar spheres. That is crystals that are less shaped like spheres burn less like spheres.

Another interesting aspect is that for given sphericities there may be multiple crystal geometries which in turn will burn up at

Table 1 Metric on the generated crystals as shown by Fig. 3.

Label	S/V	Ψ	t_{bS}/t_{bC}
(a)	13.700	0.758	0.320
(b)	22.002	0.425	0.0622
(c)	18.803	0.388	0.049
(d)	15.286	0.796	0.449
(e)	18.694	0.676	0.292
(f)	17.405	0.764	0.351

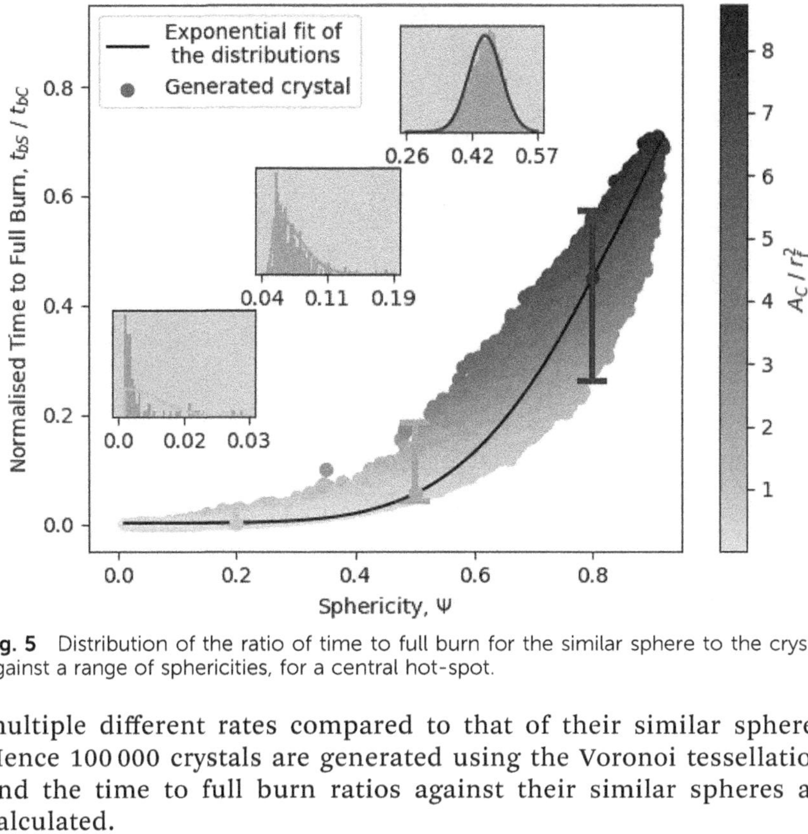

Fig. 5 Distribution of the ratio of time to full burn for the similar sphere to the crystal, against a range of sphericities, for a central hot-spot.

multiple different rates compared to that of their similar spheres. Hence 100 000 crystals are generated using the Voronoi tessellation, and the time to full burn ratios against their similar spheres are calculated.

For three chosen sphericities, a skew normal distribution is fitted to the histograms of burn time ratios, these plots show a much larger spread of values for increasing sphericities. This distribution in the burn time ratios can be accredited to the maximum distance of a crystal vertex from the crystal centre, as shown by the colour plot. Where r_f is the distance from the centre of the crystal to the furthest vertex out, *i.e.* the vertex which will be burnt up last.

Fig. 6 takes a vertical slice from Fig. 5 for a sphericity of ~0.8, there is a clear positive linear correlation between smaller values of A_C/r_f^2 (and hence larger distances to the furthest vertex from the centre) and a slower overall burn time against that of the similar sphere.

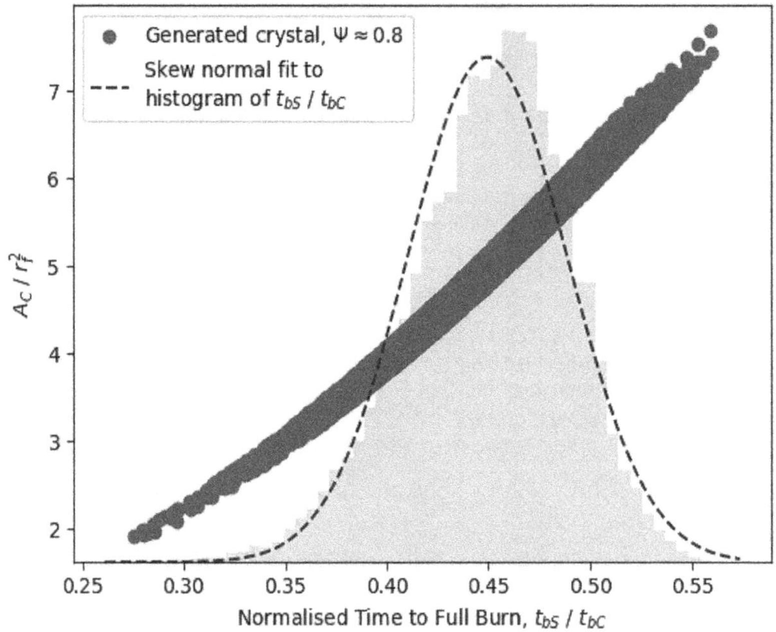

Fig. 6 The ratio of time to full burn against the metric A_C/r_f^2, for generated crystals with a sphericity of ~0.8. Along with the histogram and generated skew normal fit for the burn time ratios.

4 Edge ignited hot-spot burning

Edge ignited hot-spots are driven by inter-crystalline phenomena such as frictional heating between neighbouring crystals. The time to full burn from each of the generated crystal vertices is calculated and the average is used to normalise the time to full burn of the similar sphere, also ignited on the edge.

Fig. 7 shows the comparisons of burn fraction over time for a crystal ignited on first the closest to the centre vertex, and then the furthest from the centre vertex. Observe the difference in time to full consumption is due to the fact that for a crystal to be fully consumed by a spherical burn front starting at an edge, said burn front must have a radius equal to the furthest vertex from the starting edge. Hence depending on how the vertices are clustered, the maximum radius of the spherical burn front may be different to the furthest vertex from the centre.

Again the same 100 000 crystals are generated with the Voronoi tessellation and used to find the distribution of time to full burn ratio. This time the ratio is calculated against their similar spheres which are ignited at the edge instead of the centre. Each crystal is burned from a hot-spot starting at each of its vertices and the time to full burn is taken as an average over these times.

Again, three skew normal distributions are fitted to the same chosen sphericities, as in Fig. 5. Overall there is a much thinner spread over the range of sphericities, with a similar trend that the less spherically shaped crystals burn up much slower compared to that of their similar spheres.

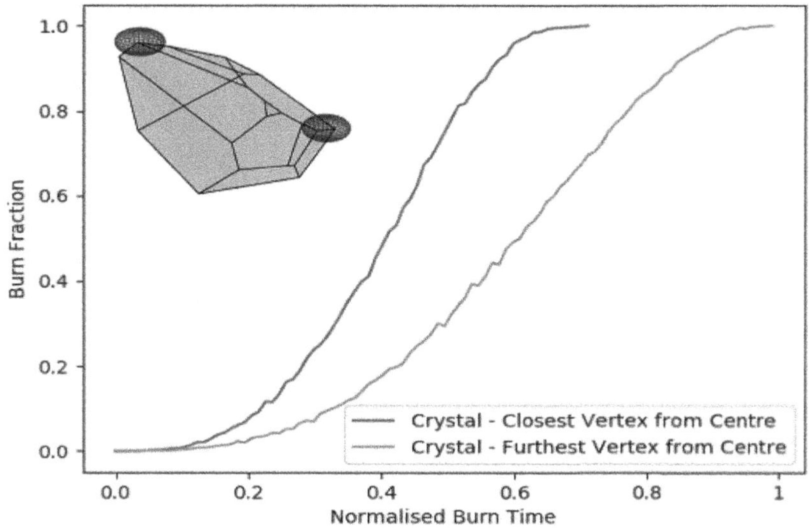

Fig. 7 Burn fractions for the edge ignited hot-spots, for both the closest and furthest vertices from the centre.

As before the colour map shows the dependence of the furthest vertex from the centre point, although the trend is slightly less prevalent in the edge ignited hot-spot case as the burn time ratios are averaged over all the vertices.

5 Conclusions

In conclusion it has been shown that over a wide range of sphericities the time to full burn consumption can be drastically slower than that of the equivalent time for the similar sphere. This is especially prevalent at the lower sphericities where the spread is thinner but the rate of burn is much slower, and at higher sphericities where the spread is much larger but there is a closer agreement with that of the similar sphere burn time.

Fig. 9 shows the combination of the bounds on the distributions taken from the centre ignited hot-spots (Fig. 5) and the edge ignited hot-spots (Fig. 8).

At lower sphericities the overall spread is low, and the means are in good agreement. Hence it can be concluded that no matter where on a crystal the hot-spot is first ignited, the time to full burn consumption will be approximately the same. It is noted that for the very low, but still realistic, sphericities that the crystals are burning up ~10 times as slow as the similar sphere equivalents.

At the higher sphericities the overall spread is much wider, with a larger discrepancy between the mean values from centre hot-spots *vs.* edge hot-spots. Meaning that there is a much larger reliance on where the hot-spot originates in the crystal for the overall time to full burn consumption. For example, a crystal with a sphericity of ~0.8 may burn up

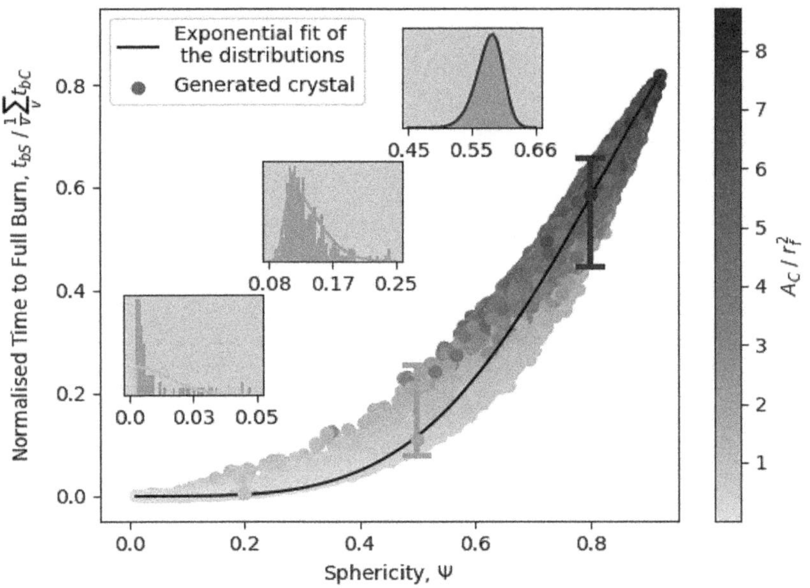

Fig. 8 Distribution of the average ratio of time to full burn for the similar sphere to the crystal, against a range of sphericities, for an edge hot-spot.

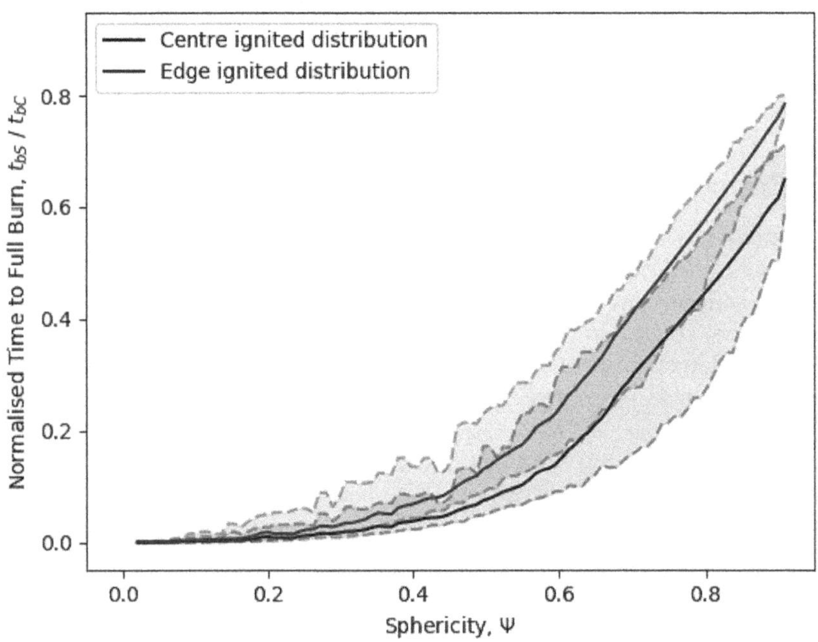

Fig. 9 Comparisons of the distributions for both centre ignited, and edge ignited burning. Along with the fitted means.

anywhere from ~1.6 times to ~4 times slower than its similar sphere equivalent, depending on where the hot-spot originated.

Future models wishing to simulate sub-sonic burning after ignition may want to implement a distribution of burn speeds to better capture this behaviour.

References

1 J. P. Curtis, J. E. Reaugh and C. A. Handley, *Proceedings of the 44th International Pyrotechnics Seminar*, 2019.

2 J. P. Curtis, A. Schofield and W. Belfield, *Proceedings of the 2nd International Explosive Conference*, 2024.

3 J. E. Reaugh, B. W. White, J. P. Curtis and H. K. Springer, *Propellants, Explos., Pyrotech.*, 2018, **43**, 703.

4 D. M. Williamson, C. R. Siviour, W. G. Proud, S. J. Palmer, R. Govier, K. Ellis, P. Blackwell and C. Leppard, *J. Appl. Phys.*, 2008, **41**, 085404.

Expansion of high-fidelity physics capabilities for simulating the detonation and behaviour of explosive materials†

A. Limbu,*[a] E. Tam,[a] M. Whittaker,[a] S. Falco,[b] D. Eakins,[b] A. Klomfass,[c] A. Tyas,[d] T. Lodge,[d] D. Farrimond[d] and D. Pope[a]

There is continuing motivation to develop and attain numerical modelling capabilities to better assess, understand and evaluate blast effects from explosive threats for UK Defence and Security. In line with this, significant efforts have been made by the Terminal Effects Simulation Team (TEST) at the Defence Science and Technology Laboratory (Dstl) to introduce novel functionalities – such as a capability to model porous explosives – which has delivered promising results, but ongoing work is required to validate outputs before advancing development. On the other hand, non-ideal explosives, including Home Made Explosives (HMEs) are an especially complex entity to understand due to various effects including metal particle combustion, non-ideal detonation and manufacturing variability. Long-term research investigations are ongoing into the methods by which the phenomena can be better understood and simulated, meanwhile, current available modelling capabilities are being explored in commercial codes such as APOLLO Blastsimulator to progress Dstl's ability to model non-ideal detonation and air blast events. This paper will detail the ever-evolving efforts to develop the numerical functionality described above, the role each capability plays in our competencies as an organisation and the future work required for further advancement.

1 Introduction

1.1 Capability gaps

There are currently some acknowledged gaps in explosives modelling that need to be rectified in order to comprehensively simulate all aspects of energetic behaviour. To address this there are ongoing efforts to improve the understanding and modelling of explosives, which includes both short-term objectives and longer-term research goals, through exploitation of existing, and development of innovative, modelling capabilities. One specific area of interest is modelling non-ideal explosive

[a] *Defence Science & Technology Laboratory, UK. E-mail: alimbu@dstl.gov.uk*
[b] *University of Oxford, UK*
[c] *Ernst Mach Institute, Germany*
[d] *University of Sheffield, UK*

detonation and blast. An explosives detonation behaviour and subsequent blast effects can be described as either ideal or non-ideal:

- Ideal – Refers to a reacting explosive with a very small reaction zone, critical diameter‡ and run-to-detonation distance§ with a near constant detonation velocity. Due to the small reaction zone, a simple programme burn model can be used to approximate the detonation behaviour,[1] where the reactant instantly transitions to its detonation products following a reaction front based on a constant detonation velocity.
  - Ideal explosives will also behave non-ideally if the charge geometry is on a small enough scale (typically a size smaller than used in reality).[2]
- Non ideal – Refers to an explosive with a larger reaction zone and critical diameter as well as a longer run-to-detonation distance with an accelerating detonation velocity. The explosive output is also dependent on the mass.[3] The larger reaction zone means that the energy release and build up to a detonation wave is slower and therefore the instantaneous transition that is assumed by a programme burn model would over predict the detonation behaviour.
  - Non ideal explosives can approximate ideal behaviour if the charge is large enough, for example ANFO detonates ideally when it has a mass of approximately >100–200 kg[2].

The non-ideal detonation behaviour is typically found in non-ideal explosives, including many HMEs. Non-ideal explosives are more challenging to model in comparison to ideal explosives for three main reasons:

- Explosive variability – Manufactured explosives, even with stringent regulations, will vary in their explosive make-up from sample to sample. This variability is far more exaggerated for HMEs as they are not manufactured with a consistent approach or regulatory controls. The constituent ingredients will vary considerably and will often include commercially available products such as fertiliser to replace laboratory-grade materials. The difference in material quality, particle sizes and the quality of manufacturing technique will naturally result in significant variation.
- Manner of energy release – Non-ideal explosives can contain other ingredients such as metal powders that change the energy release profile of the explosive. These added materials can be inert during the detonation process but will react at a later stage and combust, releasing energy over a longer timescale.[5]
- Initiation and detonation – As stated before, non-ideal explosives have larger reaction zones where the transition from reactants to detonation products occurs over a longer period than for ideal explosives.[6] Due to this large reaction zone, they often have a large critical diameter and their detonation performance can heavily depend on

‡ Critical diameter – The required diameter for an explosive for detonation to occur.[4]

§ Run to detonation distance – The distance for a shock wave to travel until it transitions into a full steady-state detonation wave for a given input pressure.[4]

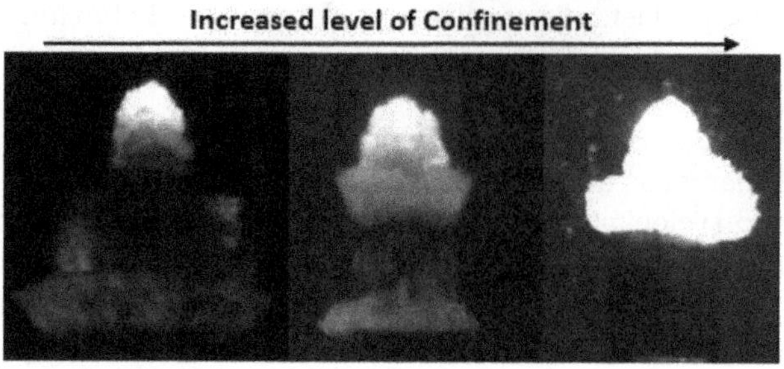

Fig. 1 Confinement study using an underwater explosive.

their level of confinement and geometry as seen in internal tests, see Fig. 1. Their detonation velocity is also typically slower, with a longer run-to-detonation distance and the detonation wave is more curved.

This paper will discuss the expansion of Dstl's high-fidelity physics functionalities for simulating aspects of the non-ideal behaviour of explosives to address these capability gaps.

1.2 APOLLO blastsimulator[7]

Each individual software option contains its own set of tools and functionalities to simulate explosives and their effects, providing the ability to model the common relevant phenomena. However, there is no single code that can accurately model everything. Therefore it is necessary to utilise multiple numerical codes, individually and/or (when possible) coupled¶ to other codes, to enable the combination of functionalities to tackle a single problem and aid in understanding certain behaviours of a given energetic material.

APOLLO Blastsimulator (referred to hereafter as APOLLO) is a high-fidelity computational fluid dynamics (CFD) code that was developed by Fraunhofer Ernst Mach Institut (EMI) in Germany. The code uses an Eulerian framework with an explicit finite volume method and has 1D and 3D solvers, with a 2D solver currently under development. It is also computationally efficient due to its dynamic mesh adaptation‖ (DMA) and zoom** functionalities.

¶ Coupling is a functionality that allows the capabilities of Euler and Lagrangian codes to be utilised in the same simulation. This can be one-way wherein the Euler code outputs the full pressure-time history as input load for the Lagrangian code, or two-way wherein the codes run simultaneously, but alternating such that the output of one feeds into the input of the other and *vice versa*.[7]

‖ Dynamic Mesh Adaptation (DMA) is a numerical capability within APOLLO which improves computational efficiency. The DMA algorithm selects at least one region, with the strongest gradient for the parameter of interest (*e.g.* pressure), resolves the region to the highest resolution specified, and applies a coarser resolution to areas with weaker gradients.[7]

** The Zoom option is used to handle increasing length scales. It allows for a time and region dependent DMA method, where a fine resolution can be initially selected and lowers over time or distance as the blast wave expands.[7]

APOLLO is an Eulerian code, which is commonly used for modelling energetic events, and it does not have the option to use a spatial Lagrangian integration scheme, which can better represent structural deformation and small amounts of distortion. Additionally, since the code does not contain strength models that are required to simulate dynamic structural response or fragmentation it is necessary to couple with a structural solver (*e.g.* LS-DYNA[††]) in such instances.

Dstl's current ecosystem includes LS-DYNA and IMPETUS AFEA with an aim of SOPHIA soon to be included, all of which possess two-way coupling capability to APOLLO. APOLLO has a library of equations of state (EOS) for a multitude of explosives, which can be used for modelling detonation and blast. Should additional EOS not present in the library be required, thermochemical codes – such as Cheetah and EXPLO5 – can be utilised to develop entries for use within APOLLO. This method has been employed for ideal explosives, such as PE4, PE8 and PE10 and extensively validated against experimental data.

A key strength of APOLLO is its ability to explicitly model afterburn (or secondary combustion), an additional combustion reaction that occurs in fuel-rich, oxygen-deficient explosives with inherently insufficient oxygen to achieve complete oxidation. In such scenarios, this then leads to partially reacted and unreacted explosive material further reacting with external oxygen sources such as the surrounding ambient air, provided sufficient energy is present to initiate the reaction.[8]

APOLLO's base functionalities allow the code to be proficient at modelling air blast scenarios. This, in addition to its ability to couple to other structural solvers to simulate material response, highlighted it as software with significant potential eventually leading to the extensive contribution to the overall development of the code. Having been extensively used and validated by the TEST team at Dstl over several years for ideal explosive air-blast simulations, it has grown to become the main high-fidelity blast code used within Dstl due to its efficiency and accuracy.

2 New high-fidelity capabilities

APOLLO's toolset has been expanded over recent years to model different explosive phenomena. These initial simulations have focused on modelling mainly ideal explosives to provide confidence in the new capabilities robustness and accuracy before modelling non-ideal explosives.

2.1 Porous functionality

Conventional physics codes are typically designed to represent explosives as homogenous materials (see Fig. 2, left), when in reality they can be composed of multiple materials (see Fig. 2, middle and right), including air in the form of pores – this is known as porosity. Dstl subsequently

[††] LS-DYNA – is a multi-physics explicit dynamics solver (based on finite element analysis) that is typically used to simulate extreme deformation of structures.[9]

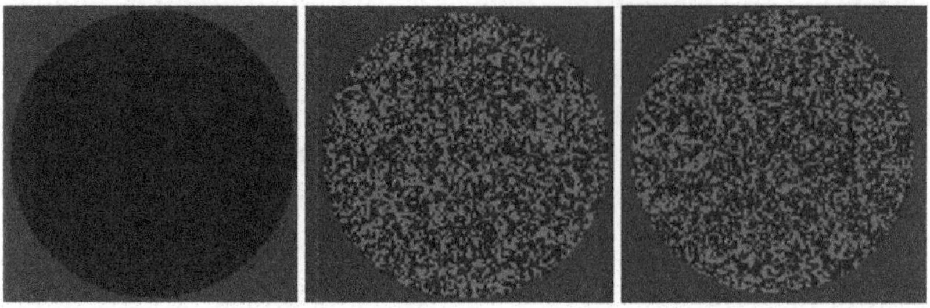

Fig. 2 Standard smeared homogenous representation of CompB (left) and stochastically-filled CompB geometries (middle and right).

decided to develop a porous functionality to enable the analysis of a more representative material.

The porous functionality works by running an executable in conjunction with a text file containing: the desired charge geometry, including shape and size; resolution; ambient conditions, for temperatures and densities; pore and volume fraction, to stochastically generate the porosity within the charge; and a seed number. A seed number is assigned to allow the charge composition to be identically recreated at a later date or changed to produce a different geometry. This information is read to generate the charge in the form of a map file, which can be read in APOLLO, and introduced to the full global model.

This functionality works robustly with APOLLO's Burn on Time (BoT) detonation model (also known as programme burn) wherein cells containing explosive material instantaneously convert from reactants to detonation products at a time calculated with the distance between the initiation point and the cell in question, whilst assuming a constant user-defined detonation velocity (see Fig. 3).[7]

Furthermore, an initial study was conducted to understand the resolution dependence of the capability; five models ranging from 100 microns to 1 mm in resolution were used to generate the charge geometry, and their blast effects were evaluated in both the near-field and the far-field. It was concluded that although there is a difference in how the explosive is physically represented (see Fig. 4), there is minimal difference in the blast effects. This indicated that this capability was not resolution sensitive in this case, however, further investigations are required for a wider range of materials and scenarios to fully understand any resolution dependence of the functionality.

2.1.1 Powdered explosives. Powdered explosives are often represented as a smeared continuum within a numerical code with a single equation of state for a specific bulk density, whereas in reality there are voids (air gaps) between the particles, which are randomly distributed. In addition to this, depending on the manufacturing quality and consistency, sometimes the bulk densities may also vary from the expected bulk density between individual charges.

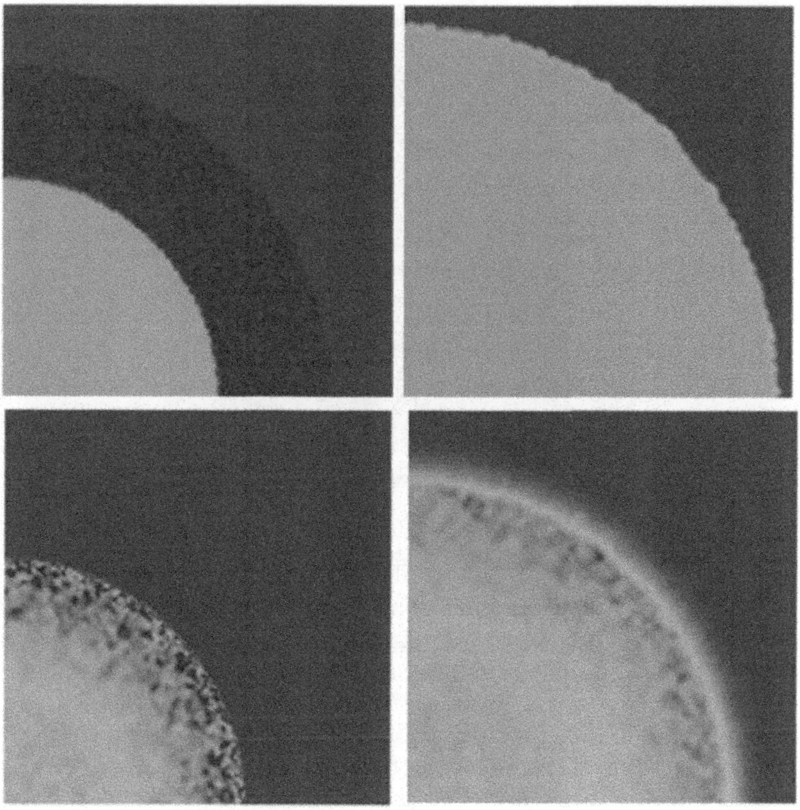

Fig. 3 Conversion from reactants to detonation products following a spherical reaction front (top) and pressure distribution of the reacting explosive (bottom).

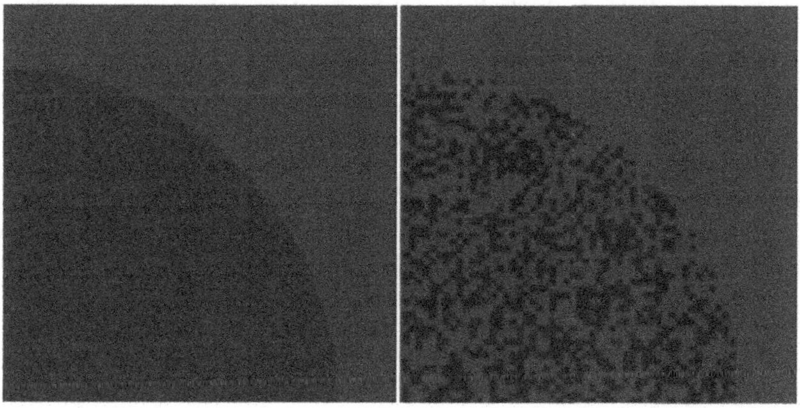

Fig. 4 Porous charge with 100 μm resolution (left) and 1 mm resolution (right).

The porous functionality allocates random pores within the powdered charge resulting in multiple charge geometries with different distributions of pores. This helps in understanding the differences and variations in blast effects based on the composition of the charge. Unlike the standard approach, which relies on a smeared continuum method and a

Fig. 5 Porous charge example in quarter symmetry.

single JWL‡‡ with a fixed initial bulk density, the porous functionality enables users to adjust the bulk density by manipulating the pore fraction within the charge. As a result, a range of charges with different bulk densities can be simulated with a single EOS and a density-dependent detonation velocity that can be derived using a thermochemical code. This eliminates the requirement for derivation of multiple EOS for a single explosive at different densities and the associated validation processes.

This capability was originally derived to model free-field tests using powdered ideal energetic materials (HMX, RDX, Octoviton and PETN). Initially, the standard JWL at bulk density was used to model the experiments, however as the bulk density recorded in the test was far lower than that applied in the associated JWL, the modelling results across the explosives were dissimilar and inconsistent to the experimental data. Therefore, the porous functionality was used, introducing pores into the charge to achieve the test density without having to derive a new JWL. Fig. 5 shows an example of one of the powdered explosive geometries generated.

This method was successful in matching the test data very well in both the near field ($Z \approx 1.25\ \mathrm{m\,kg^{-1/3}}$) and the far field ($Z \approx 6.3\ \mathrm{m\,kg^{-1/3}}$) consistently for all materials examined. As shown in Fig. 6 and 7 for one of the powdered charge materials, the APOLLO results trend very well with the experimental data and simulate similar blast effects.

2.1.2 Mixed porous explosives. The porous functionality has now been expanded to model composite explosives – explosives with multiple energetic components. However only studies on explosives without

‡‡ An equation of state is an equation that relates the pressure, temperature and volume of a substance; the Jones–Wilkins–Lee (JWL) variety is used universally to describe the expansion of detonation products, particularly in a numerical modelling context.

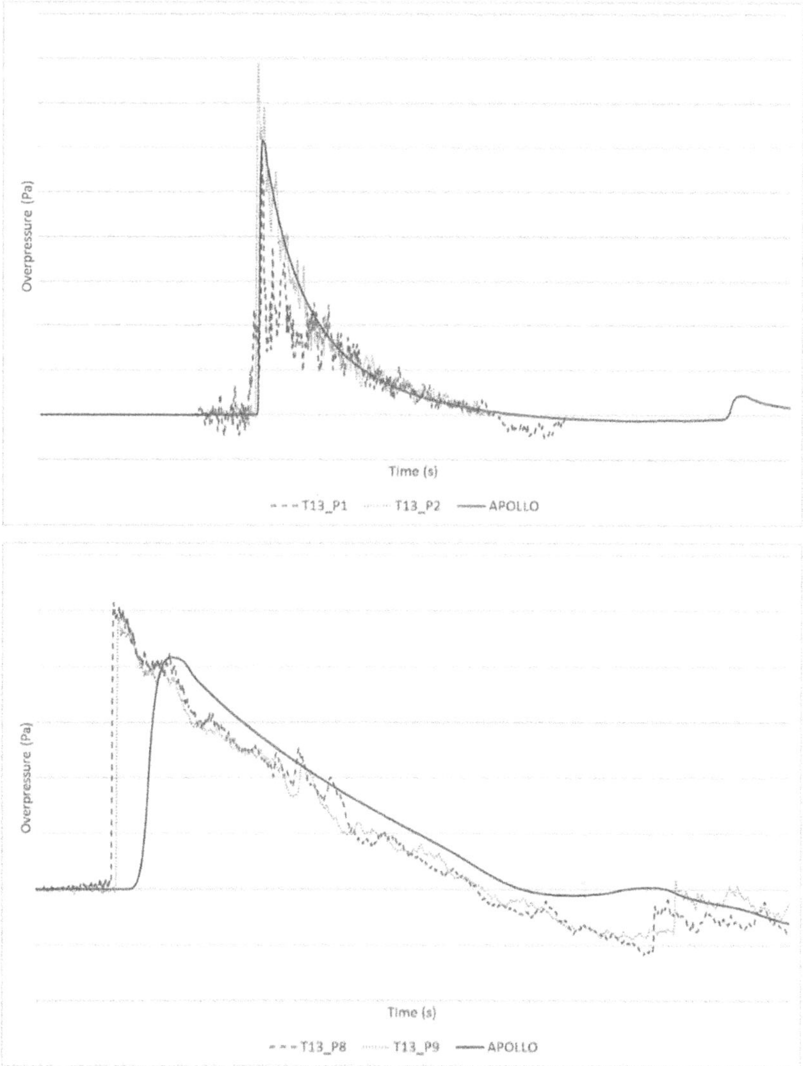

Fig. 6 Overpressure comparison in the near field ($Z \approx 1.25$ m kg$^{-1/3}$) (top) and the far field ($Z \approx 6.3$ m kg$^{-1/3}$) (bottom) for a powdered charge simulate similar blast effects.

additional binders have been conducted to date (*e.g.* CompB and Octol). In the standard approach to explosives modelling, mixed porous explosives such as CompB (≈60% RDX/40% TNT)[4] have typically been represented as a smeared continuum with a combined bulk JWL. However, with the porous functionality a CompB geometry can now be stochastically generated as an appropriately proportioned mixture of TNT and RDX. Similar to the porous charge development, the volume fraction of both TNT and RDX has to be specified to achieve the correct mass percentage of each explosive within the mixture.

To evaluate how well it can model mixed energetic explosives, five mixed TNT/RDX spheres representing CompB were stochastically

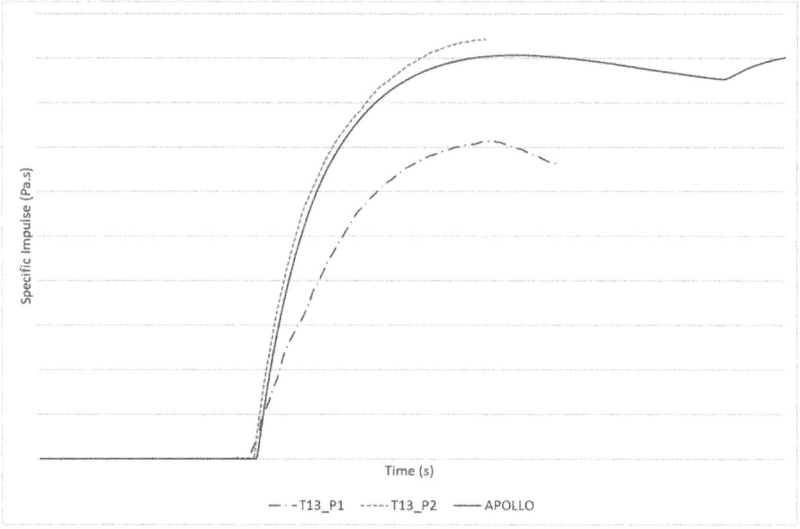

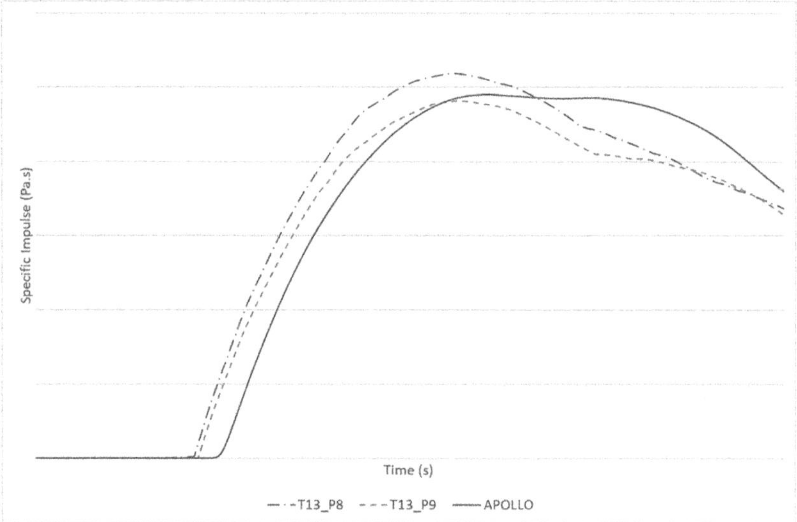

Fig. 7 Specific impulse comparison in the near field ($Z \approx 1.25$ m kg$^{-1/3}$) (top) and the far field ($Z \approx 6.3$ m kg$^{-1/3}$) (bottom) for a powdered charge.

generated (see Fig. 8) and their simulated blast effects were compared in the free field to a standard CompB JWL, where the CompB charge is modelled as a smeared continuum with a single CompB EOS.

As shown in Fig. 9, all five models show a variable output with a very fine spread in the far-field, with a larger spread in the near-field in terms of overpressure. This disparity could be due to potential localised extreme differences in the mixing. This difference is however minimal for specific impulse in both the near-field and far-field.

Fig. 10 is an extension of Fig. 9 and therefore the same conclusion can be drawn.

Furthermore, a cylinder expansion (CYLEX) test was simulated to validate the functionality. Five stochastically generated CompB mixtures

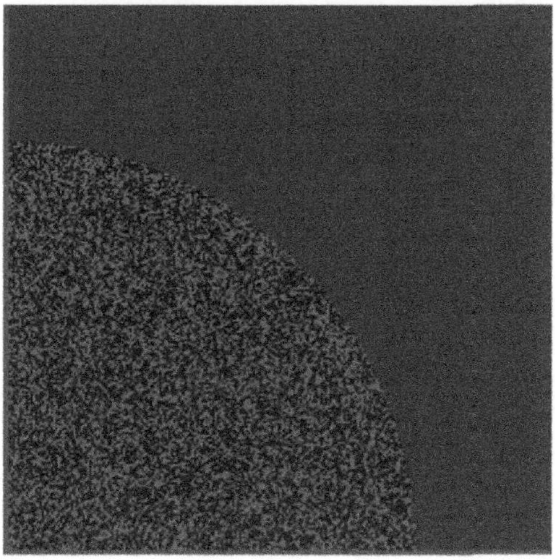

Fig. 8 CompB charge geometry example using the porous functionality in quarter symmetry.

were generated and placed as the explosive fill with a copper cylinder wall encapsulating the energetic material. The explosive is initiated from the bottom and the wall velocity of the cylinder is tracked. This model was also set up and run in Autodyn[§§] using Autodyn CompB JWL values. As seen in Fig. 11, when the CompB charge initiates and detonates, the detonation products are formed which expands the cylinder, driving it outwards.

With this study, the results of the porous functionality are compared to the standard JWL approach and experimental data, as well as Autodyn results using an EOS generated in EXPLO5.[10] In Fig. 12, both Autodyn and APOLLO outputs slightly over predict the wall velocity in comparison to the experimental data. However, when using the porous functionality the results trend very well with the Autodyn results and experimental data throughout the simulation. Both the simulation results and the comparisons show a successful application of the porous functionality to develop mixed explosives.

2.1.3 Prilled explosives and packing algorithm. There is also great interest in investigating packed prilled explosives (such as ANFO) for a given geometry. This can be achieved by pairing the porous functionality to generate the individual prills, with an algorithm to pack the prills together constructing the prilled explosive.

Two algorithms (obtained from open sources[11,12]) were incorporated into a Python script: one to perform uniform packing of equal spheres (see Fig. 13, left) and one to perform non-uniform packing of both equal (see Fig. 13, middle) and unequal (see Fig. 13, right) spheres. In both cases the script calls for a bounding radius to be prescribed as a constraint for placing spheres in three-dimensional space, the logic being

[§§] Autodyn is a commercial code developed by ANSYS US. It is a Finite Volume code and has a wide range of functionalities including simulating blast.

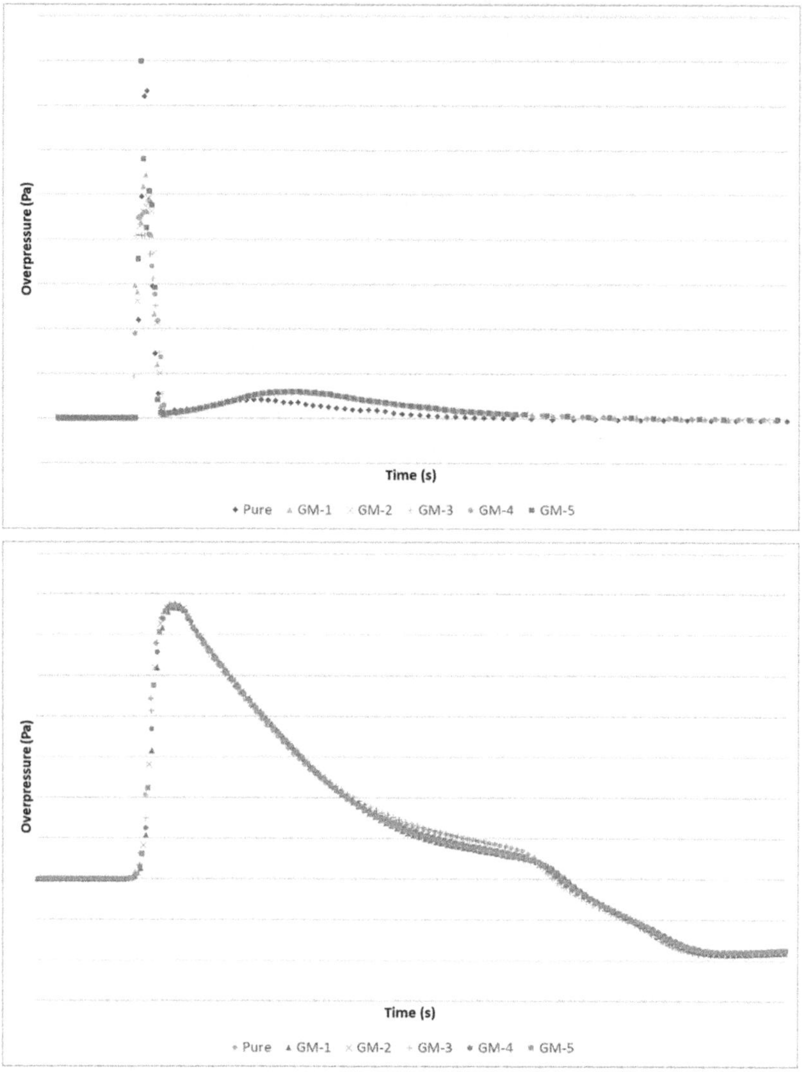

Fig. 9 CompB porous functionality comparison against a standard JWL approach taken at a scaled distance of approximately 0.7 (top) and 11 m $kg^{-1/3}$ (bottom).

that any given sphere centre, plus its radius, cannot exceed the bounding radius. It is at this point that the similarities between the two algorithms end.

The algorithm for uniform packing of equal spheres is a hexagonal close packing formula which tightly packs the spheres in ascending layers in an ABAB form, characterised by a half-radius offset horizontally relative to the prior layer that results in every other layer aligning vertically; the maximum number of layers is specified in the Python function. In contrast, the algorithm for non-uniform packing relies on a more taxing (looping) process: first, it requires an input for sphere radius (minimum and maximum), then a random coordinate is assigned to a sphere (based on the bounding radius value) before checking the

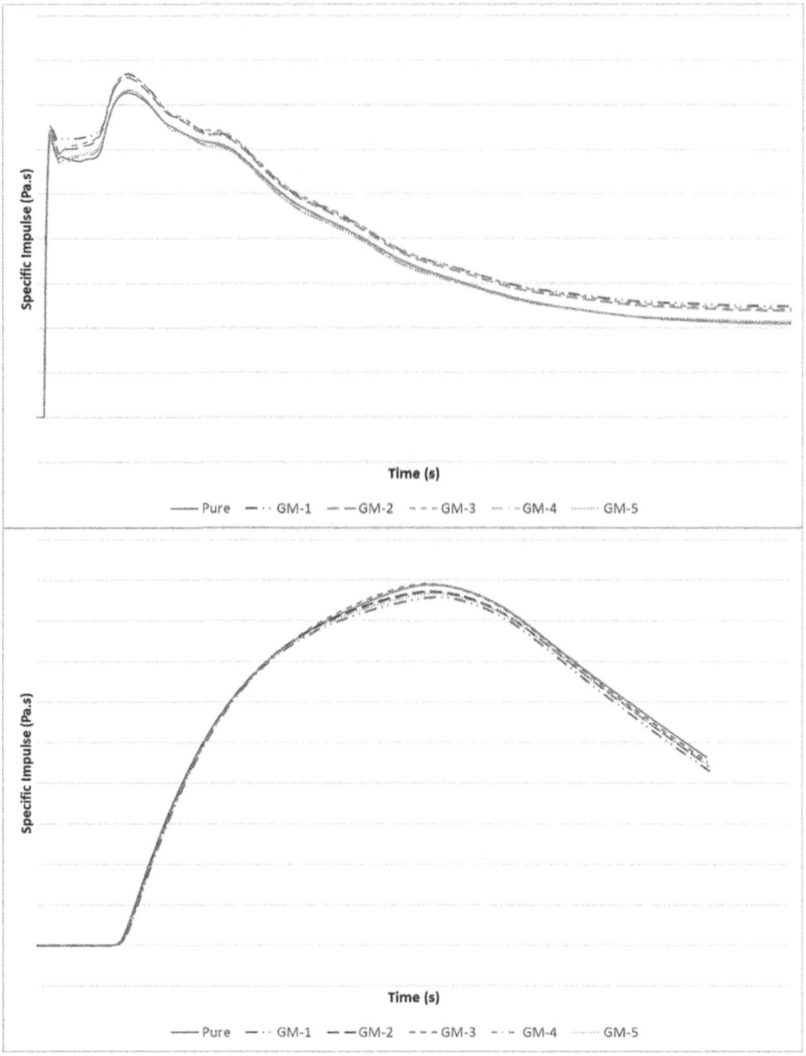

Fig. 10 CompB porous functionality comparison against a standard JWL approach taken at a scaled distance of approximately 0.7 (top) and 11 m kg$^{-1/3}$ (bottom).

Euclidean distance between existing spheres, and the bounding radius, to identify clashes. Only once all the checks are passed is a sphere accepted and added to a list that is used to check for future clashes. Given the freedom of the user to specify the minimum radius of any given sphere, two variables were incorporated to avoid excessively long (or even infinite) script runs: a maximum number of spheres and a maximum number of attempts at placing spheres, either of which will terminate the program if met.

Extensions of the conventional algorithm described above were built in to provide further flexibility in the packing; a variable was included to control the percentage of the bounding sphere to be packed to simulate a partially filled case. For example; a packing radius was also introduced forcing the algorithm to pack from the centre outwards, the idea being

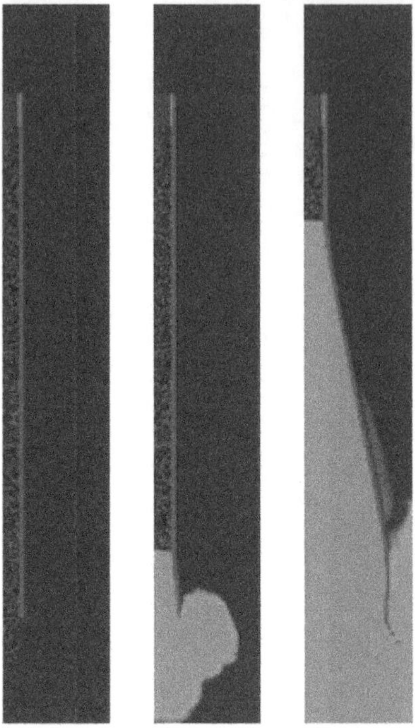

Fig. 11 CYLEX simulation using the porous functionality.

that this has the potential to better represent the effects of gravity leading to a denser pack at the bottom of a hemispherical charge (assuming packing from the effective sphere centre), for example.

Concerning the aforementioned algorithm for uniform packing of equal spheres, the above Fig. 14 illustrates the physical representation of a generic Ammonium Nitrate/Fuel Oil (AN/FO) mix.

This mix has been modelled in 1/8th symmetry and centrally detonated (see Fig. 15). The right-hand image of Fig. 15 displays evidence of imperfections in the shock front (at the extremities), although not obvious at this scale, as well as behaviour similar to spalling. Potential evidence of this has been seen in ANFO tests and is being investigated as part of a study into the detonation of ANFO. Whilst the results look promising at first glance, this is by no means considered as a valid functionality and therefore requires verification and validation prior to formal exploitation. For instance, a known issue of incorrectly represented spheres at symmetry boundaries is currently under investigation.

2.2 Reactive particles

Within APOLLO, reactive particles are Lagrangian particles that are modelled as mass points and interact with fluids through drag forces, chemical reactions and the transfer of heat. They can also interact with rigid objects in APOLLO through momentum exchange. Reactive particles can be implemented in various sizes, materials and velocities. Reactions with oxidisers are explicitly calculated and can lead to the

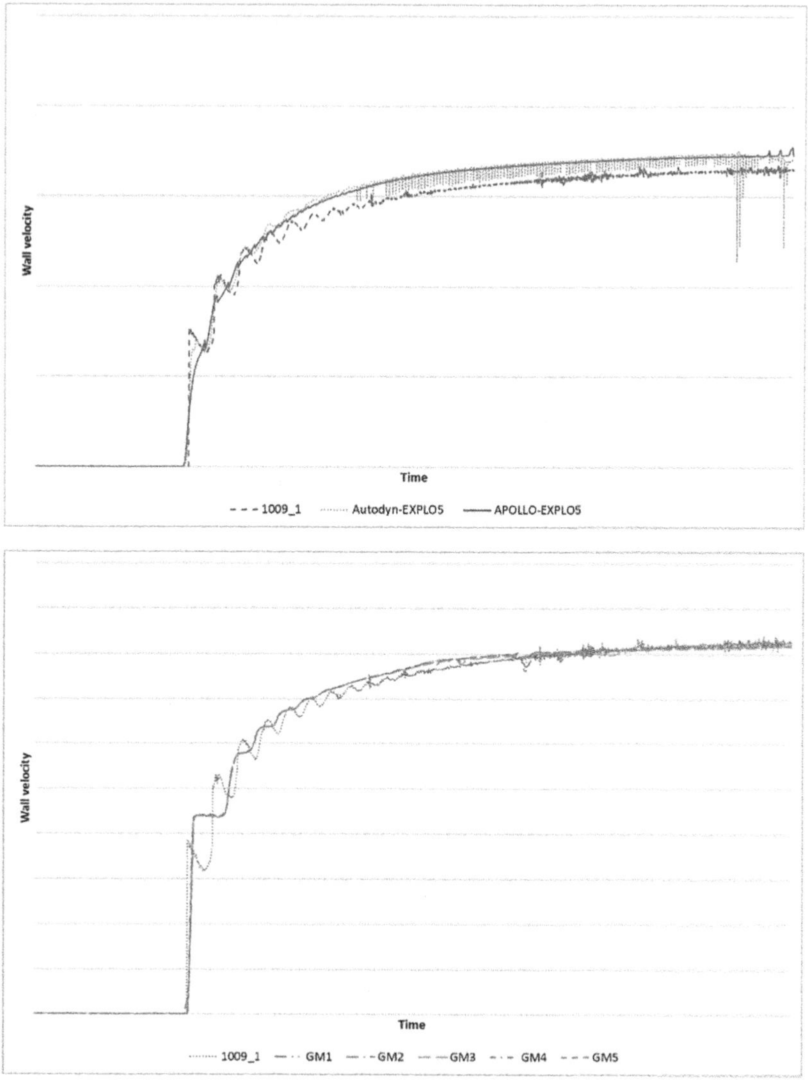

Fig. 12 CYLEX wall velocity data comparison: CompB charge single EOS comparison (top) and CompB charge cylinder expansion using porous functionality comparison with test data (bottom).

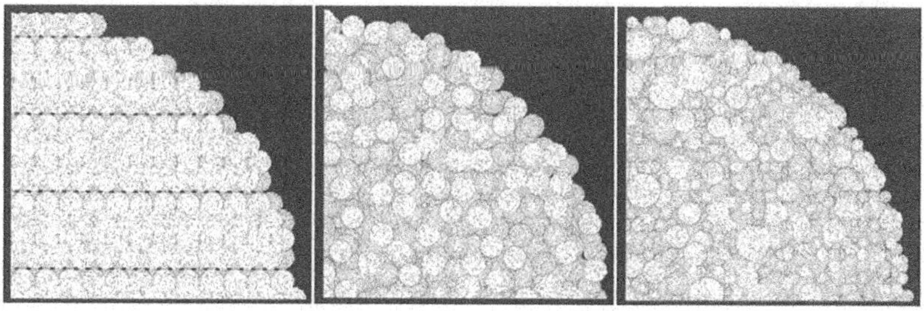

Fig. 13 Packing methods: uniform spheres fixed packing (left), uniform spheres random packing (middle), random packing of different sphere sizes (right).

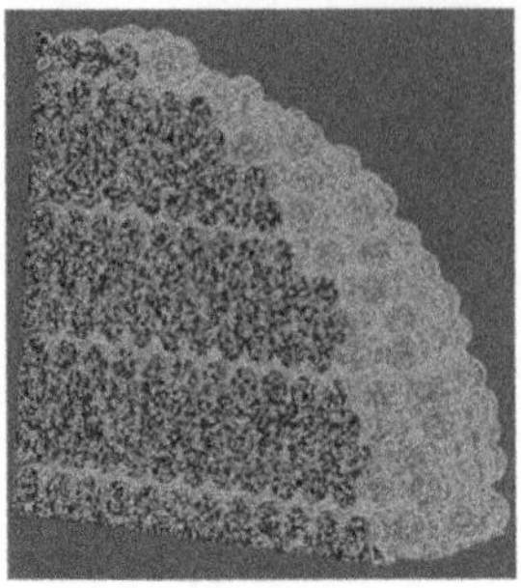

Fig. 14 Uniform spheres fixed packing of a generic AN/FO mix.

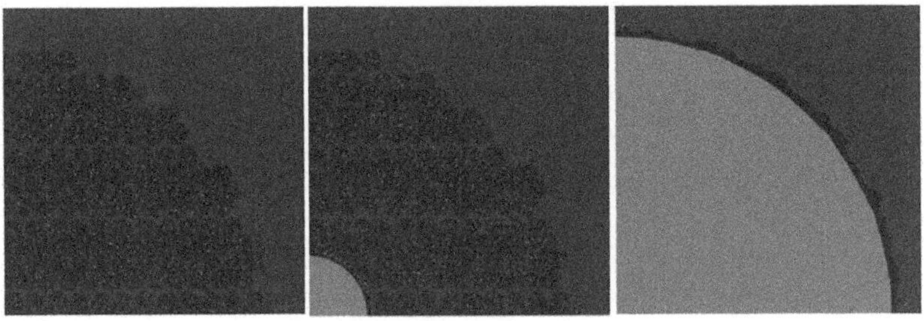

Fig. 15 Conversion from reactants to detonation products following a spherical reaction front for a prilled explosive.

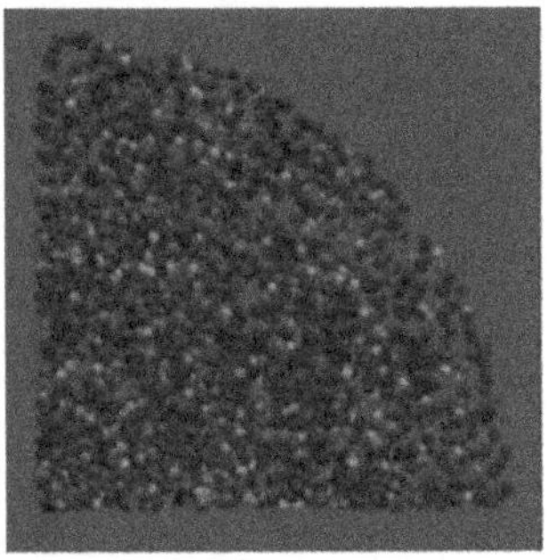

Fig. 16 PBXN-109 geometry developed using reactive particles.

formation of new combustion products such as aluminium oxide, which are modelled as a continuum condensed droplet phase.

This option is used to simulate the details of dispersion and combustion of metallic particles in high explosives interacting with a gas flow. For aluminised explosives, the aluminium powder can release a large amount of energy after detonation has occurred. Metallised explosives are modelled within APOLLO by using a continuum JWL to model the explosive as a base layer with a lower density (without the metal powder) and then reactive particles are used to model in the aluminium powder (see Fig. 16).

This function has been used to model hemispherical ground burst (see Fig. 17) and confined PBXN-109 experiments with success (the EOS

Fig. 17 PBXN-109 simulation showing the initiation and detonation of the explosive (left), the initial expansion of the blast wave (middle) and the interaction of the blast wave with the fireball (right).

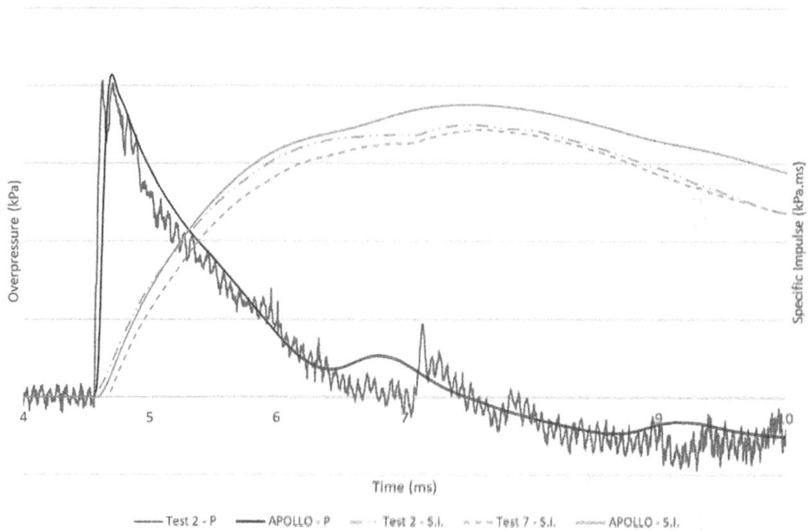

Fig. 18 250 g hemispherical ground burst PBXN-109 overpressure and specific impulse comparison.

and reaction parameters used in this study were developed in a blind study by EMI).

As seen in Fig. 18, the APOLLO results trend very nicely with the hemispherical ground burst experimental data and for the confined PBXN-109 experiments.¶¶

The APOLLO results lie between the peaks and troughs of the 50 g spherical confined PBXN-109 overpressure and specific impulse comparisons for both in air (see Fig. 19) and in nitrogen (see Fig. 20).

2.3 Implementation and verification of the history variable reactive burn (HVRB) model into a commercial code

The implementation of the HVRB model in LS-DYNA is part of a collaborative effort between the University of Oxford and Dstl. The HVRB

¶¶ Early arrival time of the secondary shock is a known general issue (not specific to these models) in APOLLO when afterburn is included, which is under investigation

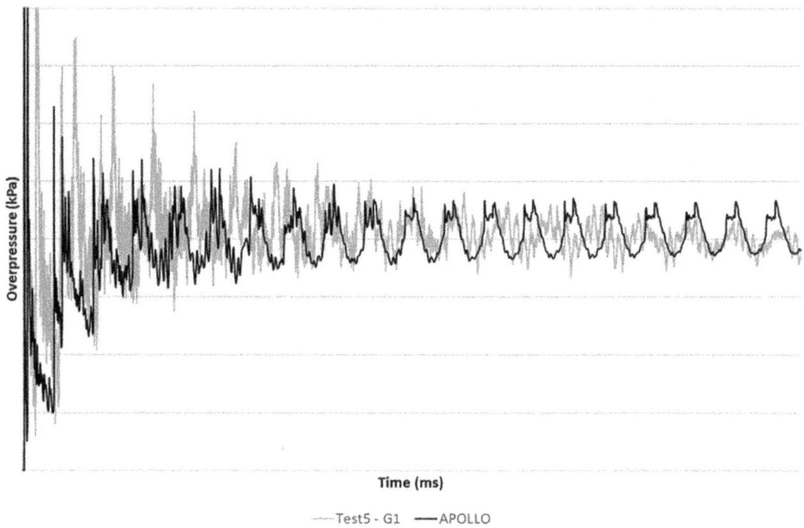

Fig. 19 50 g spherical confined PBXN-109 overpressure and specific impulse comparison in air.

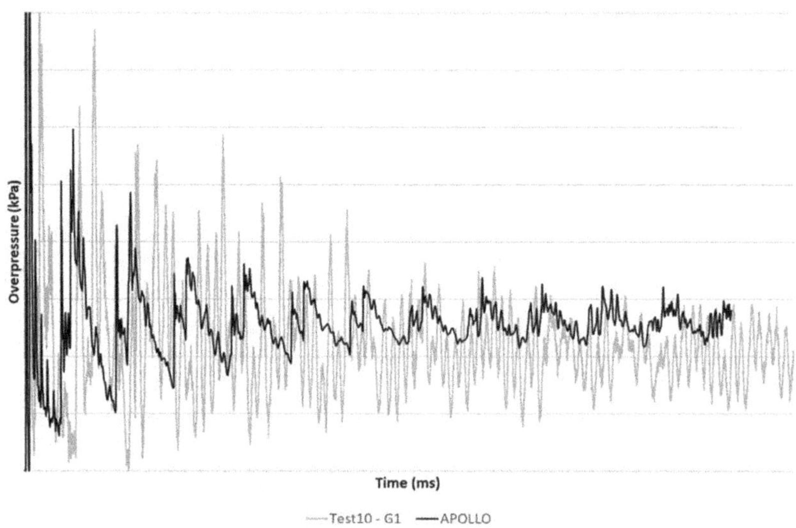

Fig. 20 50 g spherical confined PBXN-109 overpressure and specific impulse comparison in nitrogen.

model is a reactive burn model that controls the transition from the unreacted EOS to a reacted EOS. It is a pressure-based model used to treat shock-induced initiation that grows to detonation for a heterogeneous explosive material.[13]

The HVRB model was selected to be incorporated into LS-DYNA *via* user subroutine as both organisations had access to the code and it did not require source code access to implement the model, however it has been produced in a format that can be integrated into any commercial code.

To ensure that the HVRB model was correctly implemented into LS-DYNA, a verification study based on Gustavsen's flyer plate tests was

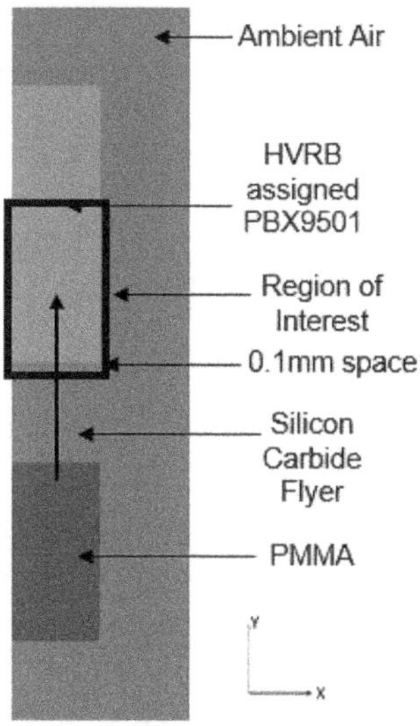

Fig. 21 2D Eulerian impact initiated PBX9501 model with a uniform fixed mesh.

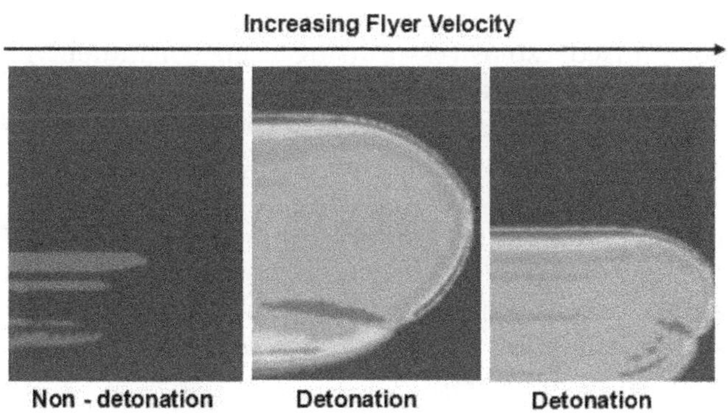

Fig. 22 HVRB model simulating non-detonation (left) and an input pressure run to detonation distance, capturing a decrease in run to detonation distance with increasing flyer plate velocity (middle and right). This occurs in the region of interest shown in Fig. 21.

conducted. This experiment was used to derive the explosive's initiation threshold and run-to-detonation distance and time, by impacting a PBX9501 explosive with flyers at varying velocities.[14]

The model was set up as a 2D axisymmetric Eulerian impact initiated explosive to match the experiment (see Fig. 21), where a silicon carbide flyer impacted a HVRB assigned PBX9501 at various velocities. It was modelled in LS-DYNA using amended HVRB values and reactant and detonation product PBX9501 EOS parameters found in online literature.

As seen in Fig. 22, within the region of interest, when the flyer impacts the explosive and provides sufficient input pressure, the explosive initiates and detonates. However, when the flyer travels at a lower velocity it does not provide the required input pressure to detonate the energetic material. Furthermore, the implemented HVRB model can capture a pressure-dependent run-to-detonation distance, where the distance is decreased with increased flyer velocity. This verification study provided confidence that the HVRB model was correctly implemented into a commercial code as it was able to capture both non-detonation and a pressure-dependent run-to-detonation distance.

3 Discussion and future work

Extensive effort and investment has gone into the understanding and the modelling on non-ideal explosives.

3.1 Porous functionality

The development of the porous functionality has been essential to address variability within explosives. The porous functionality only requires a single EOS per energetic material with a density-dependent velocity calculated using a thermochemical code and a simple programme burn model. It provides the user with a versatile tool to:

- Amend the bulk density of powdered explosives and also for mixed explosives – only requiring a single JWL per explosive material, regardless of density
- Compose mixed explosives utilising each constituent EOS, if the composite JWL is not known.
- Understand the stochastic energetic response of the explosive due to the random element-filling nature of the model

For future work, the TEST team aspires to:

- Develop the porous functionality to include binder EOS and perform subsequent validation activities – this will allow a larger range of explosives to be modelled
- Along with the packing algorithm, investigate the use of the functionality to model very heterogeneous explosives such as ANFO, explicitly modelling prills with air voids within and between prills and modelling combustion of any remaining fuel with AN detonation products to identify if realistic explosive behaviour can be modelled

3.2 Reactive particles

The addition of the reactive particle functionality allows for the reaction of metal powders such as aluminium to be modelled within a non-ideal explosive.

The initial simulations of PBXN-109 have produced excellent correlation to tests for far-field hemispherical ground bursts and confined tests with air and nitrogen atmospheres. However, further work is required to investigate:

- The effect of particle size distribution
- Near-field validation

- Previous work conducted with EMI
  - This has shown aluminium reaction with detonation products and potentially nitrogen atmospheres which did not appear to occur within the tests conducted by Dstl. Additional efforts will be made to understand when these reactions occur and any temperature or similar requirements
- Alternative metallised explosive, to ensure consistent performance of functionality

If this approach is successful further efforts will be made to extend the functionality to be used for:

- Reactive materials – coupled to structural software to simulate the fragmentation and following combustion of reactive metals (such as aluminium casings)
- Underwater Explosive (UNDEX) – many explosives designed for use underwater (*e.g.* PBXN-111) not only contain metal particles for fuel but also oxidiser particles to provide the oxygen for the fuel to combust (as there is no ambient oxidiser)

3.3 Implementation and verification of the history variable reactive burn (HVRB) model into a commercial code

The successful verification of the HVRB model implementation provides confidence that the reactive burn model can be incorporated into other commercial codes such as APOLLO. The inclusion of a reactive burn model will allow a code to better model the complex initiation and detonation of an explosive, rather than relying on a programme burn model, which does not consider any initiation or detonation criteria and assumes a perfect detonation. While the HVRB model is primarily used for ideal explosives, it represents a significant improvement from a programme burn model and allows for, in its standard form, the run-to-detonation to be modelled.

As the HVRB model input parameters are resolution and code dependent, an optimisation tool to derive HVRB parameters from experimental pop-plot data is currently being developed. The optimisation tool works by:

- Providing the tool with a target pop plot and velocity data and starting HVRB parameters
- Shock to detonation models are run at a certain resolution at multiple velocities and the pop plot is then extracted from the simulations
- This generated pop plot and velocity data is then compared to the target pop plot and velocity data and if it does not match, the HVRB parameters are readjusted
- The optimisation procedure continues until the target pop plot is recreated

Once the optimisation tool is fully developed, there is interest in integrating the HVRB model into APOLLO Blastsimulator in the future.

4 Conclusions

The TEST team have been working determinedly to attain and develop new and exciting numerical modelling capabilities to simulate aspects of non-ideal blast to fulfil pressing capability gaps. Whilst some functionalities are still in their early stages, there is ongoing and future planned work to validate, improve and exploit these numerical tools. This will involve further research, investment and further extensive collaboration with other government departments, academia and industry.

Acknowledgements

We would like to thank, in no particular order, the University of Oxford, the University of Sheffield and Fraunhofer Ernst Mach Institut for their tremendous efforts in supporting our work through their extensive knowledge and guidance. The TEST team would also like to thank the late Prof. Dan Pope, a champion of defence development, relentless civil servant, esteemed colleague, mentor to many and friend to many more. We cannot thank him enough for the influence he has had on the industry, the people, and the shaping of our work.

References

1 D. G. Farrimond, *Characterisation of Blast Loading from Ideal and Non-ideal Explosives*, PhD thesis, University of Sheffield, 2023.

2 O. Sandus, *Concepts of Ideal and Non-ideal Explosives*, US Army Armament Research and Development Command, 1981, ARLCD-TR-810421981.

3 J. Petes, R. Miller and F. McMullan, *User's Guide and History of ANFO as a Nuclear Weapons Effect Simulation Explosive*, Kaman Tempo, KT-83-012 (R), 1983.

4 P. W. Cooper, *Explosives Engineering*, Wiley-VCH, Germany, 1st edn, 1996.

5 J. Dong, W. Wang, X. Wang, Q. Zhou, R. Miao, M. Du, B. Tan, Y. Wang, T. Zhang, Y. Li and F. Cao, Effects of Nano Aluminium Powder on the Mechanical Sensitivity of RDX-Based Explosives, *Nanomaterials*, 2021, **11**(9), 2182.

6 R. S. Janesheski, L. J. Groven, S. Son, presented in part at Shock Compression of Condensed Matter, Chicago, June 2011.

7 A. Klomfass, *APOLLO Blastsimulator Manual* (Version 2024.1), 2024.

8 L. Donahue, F. Zhang and R. C. Ripley, Numerical Models for Afterburning of TNT Detonation Products in Air, *Shock Waves*, 2013, **23**(6), 559.

9 Ansys LS-DYNA Multiphysics Solver, https://www.ansys.com/en-gb/products/structures/ansys-ls-dyna, (accessed July 2024).

10 EXPLO5, *Version V6.05 User Guide*, 2020.

11 T. Hobbs, A Randomised Approach to Circle Packing, https://www.tylerxhobbs.com/words/a-randomized-approach-to-circle-packing (accessed November 2023).

12 Stack overflow, Packing Hard Spheres in a Box, https://stackoverflow.com/questions/65192816/packing-hard-spheres-in-a-box (accessed November 2023).

13 J. Starkenberg and T. M. Dorsey, *An Assessment of the Performance of the History Variable Reactive Burn Explosive Initiation Model in the CTH Code*, Army Research Lab Aberdeen Proving Ground MD Weapons Technology Directorate, ARL-TR-1667, 1998.
14 R. L. Gustavsen, L. G. Hill, R. R. Alcon and S. A. Sheffield, *Shock Initiation of New and Aged PBX-9501 Measured with Embedded Electromagnetic Particle Velocity Gauges*, LA-13634-MS, 1999.

TATB thermal decomposition: an improved kinetic model for explosive safety analysis

Keith R. Coffee,*[a] Jason S. Moore,[a] Ana Racoveanu,[b] Keith D. Morrison,[a] Alan K. Burnham,[c] Batikan Koroglu,[a] Gregory L. Klunder[a] and John G. Reynolds[a]

TATB (1,3,5-triamino-2,4,6-trinitro benzene) and polymeric formulations, are widely used energetic material applications. LLNL, as with many other laboratories, evaluates these materials to assure safe handling. This is a summary of the process of developing a new, expanded thermal decomposition model to address the behavior of TATB and formulations exposed to abnormal thermal environments. Ultimately, this model is used to assist determining the safety aspects of TATB, or any other explosive, when thermally compromised. This work is the culmination of global behavior and molecular characterization studies to increase the accuracy of the reaction sequences in the model. This report will review the science developed leading to the model, culminating with the model.

1 Introduction

TATB is the basis of all the insensitive high explosives (IHEs) of today. The compound itself has been around since the later 1800s, but it was not until the second half of the 1900s that it was recognized as an explosive. Fig. 1 left shows the molecule itself, having the carbon configured into a benzene ring, with alternating amino and nitro substitutions around the ring. It is fully substituted, with all positions on the benzene ring occupied, and these functional groups available for secondary interactions of hydrogen bonding, both intra and intermolecular. Fig. 1 middle shows distorted planar array of TATB molecules linked by in plane hydrogen bonds. These hydrogen bonds extend in three dimensions, which maintains the somewhat distorted structure. Fig. 1 right shows how these planes come together as layers forming a tertiary structure.[1–3]

TATB is considered an insensitive explosive because of the lack of response to external stimuli, such as impact, friction, spark and heating. To put it in perspective, the decomposition temperatures are easily 100 °C more stable than PETN or HMX. Because of the wide use of TATB, its thermal stability is an important issue. The impetus of this research is to understand what happens to TATB when heated or exposed to an abnormal thermal environment, whether purposely or not. The development of a thermal decomposition model is the major driving force because it is related to safety.

[a] *Lawrence Livermore National Laboratory, Livermore CA, USA. E-mail: coffee3@llnl.gov*
[b] *NorthWind Services, LLC, Livermore CA, USA*
[c] *Stratify//MH Chew Associates, Livermore CA, USA*

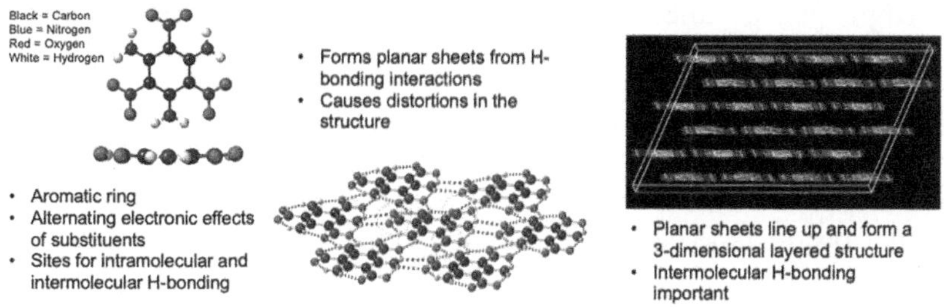

Fig. 1 TATB structure: left side—single molecule, middle—single layer array, right side—tertiary layering.

Fig. 2 Experimental examples of laboratory configurations for thermal treatment of TATB materials.

Fig. 2 shows the results of laboratory experiments designed to explore the effects of thermal exposure at various severities.[4] Under mild conditions,[5] the top of Fig. 2, TATB can go from discoloration (200 °C/6 h-image) to gradually attrition (250 °C/6 h-image). In more severe conditions[6] (SSVCT images), the heating can lead to char formation and enough gas production to open confinement. If heated under severe conditions[7] (STEX images), the reactor will rupture, scattering the char and unreacted TATB.

The model was built from earlier work proposing a reduced reaction network.[8] It uses only explicit chemical reactions. The data were used to fit rate parameters for proposed reaction schemes in a MATLAB thermos-chemical computational model. These parameterizations were carried out utilizing a genetic algorithm optimized on LLNL's high performance computing clusters, which enabled significant parallelization including a multi-step decomposition model, identification of likely autocatalytic gas-phase species, predictive sensitization and effects of confinement.[9]

Five key areas of characterization research helped in the development of the model:

- Heat and mass balance to develop global kinetic models (Differential scanning calorimetry and thermogravimetric (DSC–TGA) analyses)
- Pyrolysis decomposition with on-line analyses to detect gaseous decomposition compounds (pyrolysis-chromatography-mass spectrometry (py-GC–MS))

- High performance chromatography to separate decomposition compounds in the solid phase with molecular identification by mass spectrometry (DMSO extraction with HPLC-DAD/MS)
- Synthetic substitution of spectroscopic probes into TATB structure (isotopic substitution, ^{15}N, ^{2}H, ^{13}C, ^{18}O)
- Characterization by miscellaneous spectroscopic methods of thermal products (SS NMR, FTIR, hot stage microscope with IR, XPS, TEM, PXRD)

Characterization on both the global-level and molecular-level was key to development of the model. The global experimentation gave us the overall behaviour and the molecular level delineated the specific reaction chemistry occurring.

This report is derived directly from the presentation (Session 3: AI & Machine Learning) at the 2024 International Explosives Conference in Edinburgh, Scotland, and summarizes the development of a new and improved thermal decomposition model for TATB.[10] More complete details of the experiments, data, and model developments are given in the references within.

2 Development of model

2.1 Background in TATB thermal decomposition models

Fig. 3 summarizes of the thermal decomposition models for TATB, starting around 1950. The challenge for this type of chemistry is how to evaluate how solid-state materials begin to react or decompose. The first models had single-step reactions of reactants converting to products, such as the Prout–Tompkins.[11] As models progressed in complexity due to more experimental input, they have increased in steps of reactions, both parallel[12] and in series.[13] An important addition was earlier in this century specifically calling out the involvement water in the first step.[14] This is relevant when considering at the same time, many studies

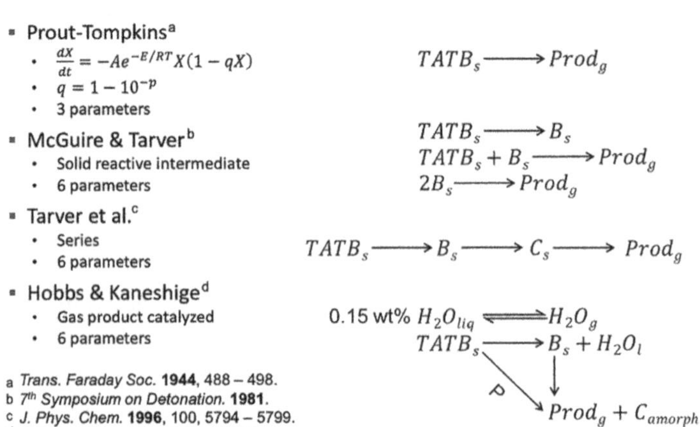

Fig. 3 Development of TATB thermal decomposition model from the literature.

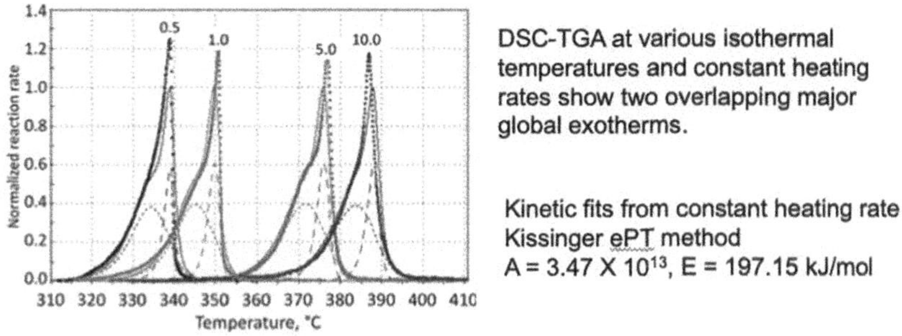

Fig. 4 DSC profiles of TATB at of 0.5, 1.0, 5.0, and 120 °C min^{-1} heating rates.

observed the formation of furazans as a first thermal reaction steps and by extension, the condensation reaction forms water.

2.2 TATB global thermal profiles

Fig. 4 shows the results of the differential scanning calorimetry (DSC) examination of TATB. These results were generated by measuring heat flow and mass loss on analytical equipment (simultaneous DSC and thermogravimetric analyses, TGA). The profiles are from multiple heating rates (0.5, 1.0, 5.0, & 10.0 °C min^{-1}). Although the range of temperatures of response varies depending upon the heating rate, the overall profile shape remains reasonably consistent. The profiles can be resolved into two overlapping exotherms with the leading-edge shoulder overlapping with a sharp major exotherm. These represent two global reactions, and with the proper examination by DSC and TGA, the global kinetics can be derived. The application of kinetic analysis using different constant heating rate profiles and a typical kinetic analyses method (in this case Kissinger's method) the activation energy parameters can be derived.[15]

Earlier work using mass spectrometry (MS) demonstrated the evolving gases resulting from thermal decomposition can be different indicating the precursors to these two global reactions are not the same.[16] Several studies have verified the composition of the volatile species evolving change with decomposition temperature.

2.3 Compound evolution during DSC

Recent efforts have analysed the gaseous effluent during thermal decomposition in more detail. Utilizing MS, the molecular composition of volatile compounds can be measured. Fig. 5 shows two examples such evolving compounds. The left side shows compounds evolving from a large-scale cook-off.[17] The experiment was heated at 29 °C min^{-1} heating rate. The evolution shows some gases evolve sooner than others and before the main event which starts around 236 °C. Fig. 5 right side shows a simultaneous DSC TGA coupled to MS (SDT–MS) experiment monitoring light gases at a 1 °C min^{-1} heating rate.[18] This experiment also correlates the gas evolution with the exotherms, as seen in Fig. 4. Again, there are two sets of evolving gases correlating with the two global reactions. It is notable that CO_2 and H_2O are early evolving and stand out.

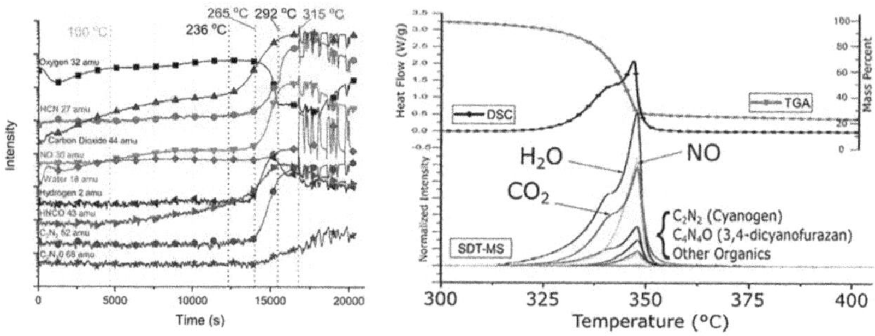

Fig. 5 Evolution profiles: left side—masses evolving from monitoring of large-scale cook-off of TATB; right side—heat flow, weight loss and masses evolving from the SDT–MS of TATB at 1 °C min^{-1} heating rate.

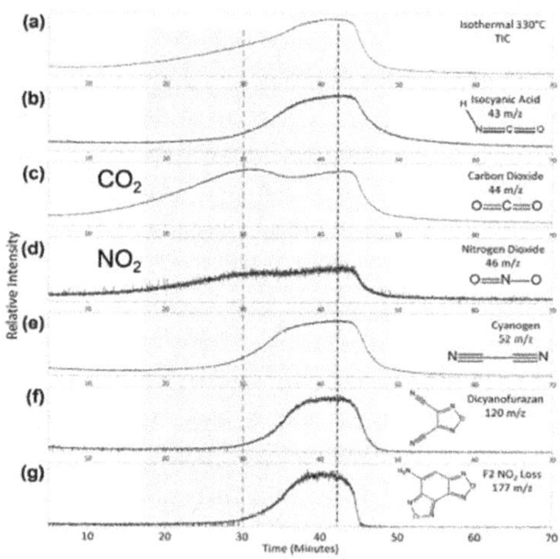

Fig. 6 Pyrolysis GC/MS of TATB heated isothermally at 330 °C for 70 min.

2.4 Compounds evolving during pyrolysis mass spectrometry

Further identification of the compounds evolving during thermal degradation comes from flash pyrolysis of TATB with MS detection in isothermally heated samples.[19] Fig. 6 shows an example of the evolution profiles at 330 °C over a period of 70 min. The top profile is the total ion current (TIC) and below are the behaviors of selected individual molecules during heating. The bimodal distribution seen in the constant heating rate experiments, as in Fig. 4 and 5, is not pronounced in the TIC profile. However, CO_2 and NO_2, and H_2O evolve early (the dashed red line) while all the materials evolve later (the dashed purple line). The CO_2 comes from the oxidation of the TATB ring. This has been proven by decomposing ^{13}C and ^{18}O isotopic substituted TATB compounds under the same conditions. The role of NO_2 also appears to be an early forming reactant and offers a new pathway for TATB decomposition.

2.5 Decomposition pathways of TATB—global and molecular

Establishing the relationship of the overall reaction behavior with the molecular composition yields insight into molecular origins of the two

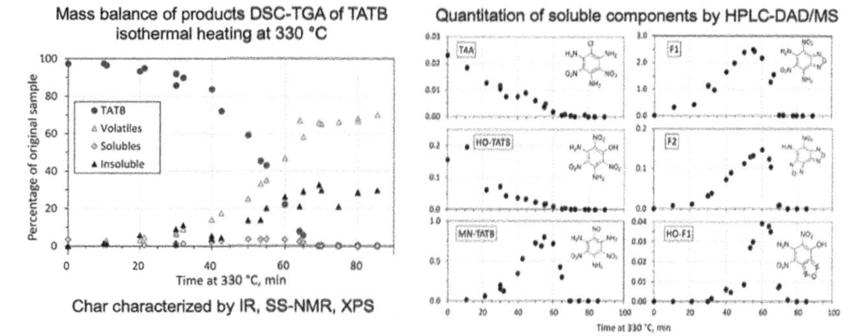

Fig. 7 Left side: mass balance of products from the DSC–TGA of TATB heated isothermally at 330 °C for 90 min; right side: selected molecular species identified from products in the DSC–TGA experiments.

global overlapping exotherms. Fig. 5 and 6 indicate the precursor compounds to the evolved species for the two exotherms are possibly different. TATB was isothermally treated in a DSC–TGA for different reaction times producing a suite of samples. The heat flow and weight loss observed and sample masses before and after were measured to provide a pseudo-mass balance.[20] The molecular composition at each reaction time was also measured by extraction of the soluble portion followed by compositional analyses.[21,22] Linking the molecular profile with the mass distribution helps correlate global and molecular reactions. Fig. 7 left side shows the residue composition divided into soluble (DMSO extractable components), insoluble (DMSO insoluble residue) portion, volatiles (mass difference before and after reaction), and the TATB concentration (by analyses of the extractable portion). These results were obtained from small samples painstakingly handled to give a mass balance and analytically accurate composition. TATB decreases in a sigmoidal behavior and is no longer present at 65 min. The solid and volatiles gradually grow in concentration. However, at 65 min, there appears to be a disruption as the TATB concentration disappears (as measured by extraction). the volatile and solid content increases.

The same samples were extracted with DMSO, and the composition determined by column separation followed mass spectrometric identification.[21] TATB was quantitated by methods recently developed.[22] At 65 min, there is a real change in the composition, which is when the TATB becomes exhausted. However, volatiles and char continued to be produced. Notice the sigmodal behavior of TATB decomposition. Fig. 7 right side shows some of the other species that were found in the soluble extract by HPLC/optical/MS detection developed in this project. For the other species T4A and HO-TATB, which were found in the starting material, monotonically decrease with time. Reaction products related to TATB, the furazans (F_1, F_2, HO-F_2) and mono-nitroso-TATB (MN-TATB) obviously were formed then destroyed during the heating. These also had the concentration drop-off at 65 min. These results elucidate the benzo-furazan decomposition pathways.

3 Chemical reaction pathways

The above provide examples of global and chemical characterization of the thermal decomposition pathways of TATB. Through molecular characterization techniques developed for this project, these results have produced two overall pathways for thermal decomposition—furazan and oxidation by NO_2. Fig. 8 shows the general types of reactions.

The furazan pathway on the left side has been known for several decades and has been well studied.[6,16,17,23,24] This current work has added some more decomposition products and conditions.[19] The other pathway on the right side, which has been recognized for many explosives, but not for TATB, is the oxidation mechanism.[18] NO_2 is involved, but it is not clear whether it is the active species or reacts producing active species. However, this has not been proposed or verified for TATB previously. These two major characterization schemes have greatly expanded the model to be more inclusive to chemistry and more accurate for prediction.

Fig. 9 shows more details of these chemical reactions and the inter relationships among the chemical species, and includes some of the

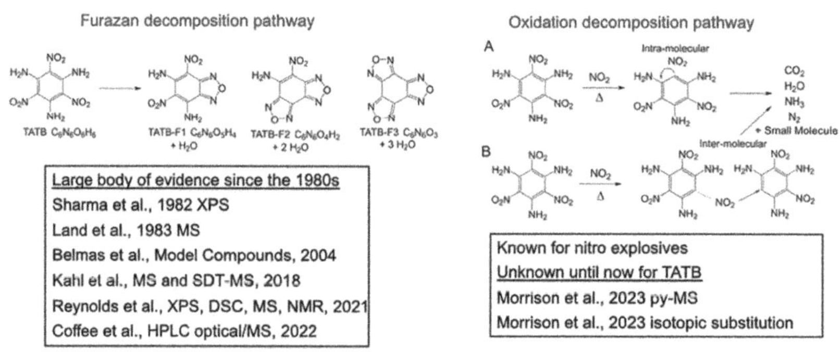

Fig. 8 Summary of chemical reaction for the thermal decomposition pathways of TATB.

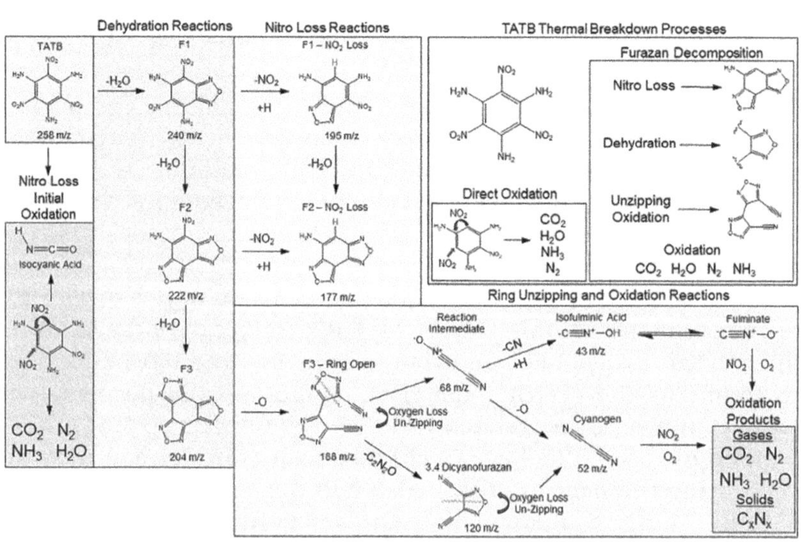

Fig. 9 Detailed chemical reaction network for the thermal decomposition pathways of TATB.

species from the characterization work.[25] The NO_2, a fierce oxidant, can come from various species. Also, this shows how the ring unzips producing species like C_2N_2 and HNCO (bottom right).

4 Improved thermal decomposition model

4.1 Model description

Fig. 10 shows the improved thermal model, expanding the number and complexity of the reactions compared to the earlier models cited above.[9] This model includes more detailed analyses of the intermediates and products both on the bulk and molecular level. It is important to point out the roles of the intermediates molecularly identified for the first time. Chemical species, such as NO_2 and early forming CO_2, are recognized as critical components. Briefly, the TATB reacts to give mono-benzo-furazan (F_1). The F_1 continues to react forming more condensed products called solid intermediates or SIs. These are a variety of compounds molecularly identified and some examples are discussed in preceding sections. Also, TATB loses and reacts with early forming NO_2, something not delineated until now, which causes oxidation reactions. NO_2 can be produced by scission from other species as well, such as dibenzo-furazan (F_2). See Fig. 9 for other examples. Many light gases are formed, such as HNCO, C_2N_2 *etc.*, with the final products being a carbon-nitride residue (CN residue) and several non-condensable gases.

Fig. 11 shows the reaction network.[9] A genetic algorithm was used to parameters orthogonalized E_a and k_0 for five reactions. The first reaction is the water equilibrium step. Direct formation of mono-benzo-furazan leads to water formation as well. Mono-benzo-furazan decomposes to solid intermediates which include other furazans, as indicated in Fig. 9. TATB also reacts with some form of NO_2. F_1 also reacts with some form of NO_2. The solid intermediates react giving char residue. The is a solid residue reaction that has the C_2N_2 term dependence.

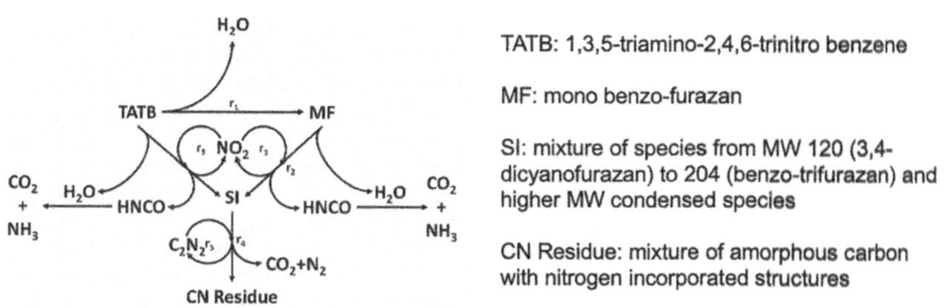

Fig. 10 Improved thermal decomposition model for TATB.

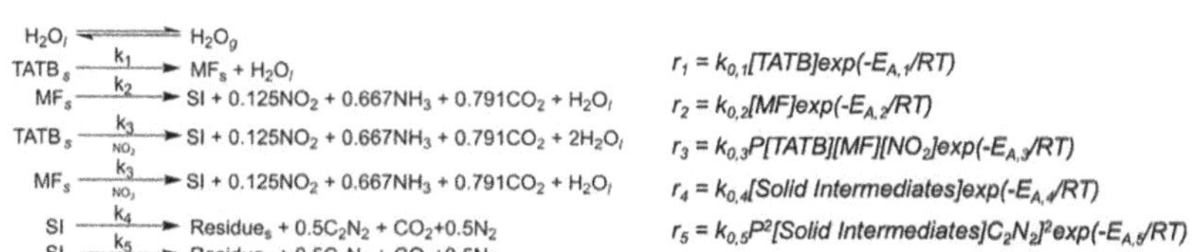

Fig. 11 Reaction network and rate constants for improved TATB thermal decomposition model.

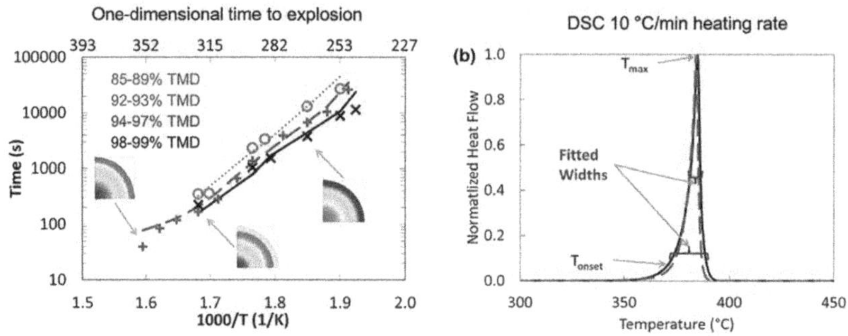

Fig. 12 Examples of model simulation compared to experimental data for improved TATB thermal decomposition model.

4.2 Examples of simulations from the model compared to measurements

The model was developed with a variety of data including one dimensional time to explosion (PODTX), DSC, TGA, burst time, *etc.*[9,26] Fig. 12 left side shows the comparison of the modeled PODTX for LX-17 (92.5 wt% TATB and 7.5 wt% Kel-F) indicating times to explosion at four different densities showing the effect of gas pressure on the rate of autocatalysis. The insets are the temperature gradients in the residual material at the end of the simulations. Heat transfer varies depending upon temperature. Fig. 12 right side shows a comparison of the DSC data and simulation at $10\ ^{\circ}\mathrm{C\ min^{-1}}$ heating rate. The overlapping exotherms shown in Fig. 4 and 5 are better seen in simulations under isothermal decomposition conditions. More detailed comparisons, such as TGA, pressure profiles in PODTX, and explosion/burst times are provided elsewhere.[9,26]

5 Conclusions

A new, improved thermal decomposition model for TATB and related materials has been developed which includes improvements in number and types of reactions. Comparison of simulations with thermal decomposition experimental data show excellent agreement.

Global reaction and molecular characterization studies provided new information relevant to development of detailed features of the model.

- Global reaction network must include at least two autocatalytic reactions, either in parallel or series, and at least one must have an explicit initiation reaction having a low activation energy
- Molecular characterization shows two predominant decomposition reactions—furazan decomposition and oxidation facilitated by NO_2
- Ring carbon oxidation occurs early in competition of furazan condensation reactions

This work addresses the need for an improved safety understanding when dealing with TATB and formulations exposed to abnormal thermal environments.

Acknowledgements

This work was performed under the auspices of the U.S. Department of Energy by Lawrence Livermore National Laboratory under Contract DE-AC52-07NA27344.

References

1 H. H. Cady and A. C. Larson, *Acta Crystallogr.*, 1965, **18**, 485–496.
2 Z. Chua, C. G. Gianopoulos, B. Zarychta, E. A. Zhurova, V. V. Zhurov and A. A. Pinkerton, *Cryst. Growth Des.*, 2017, **17**, 5200–5207.
3 H. E. Mason, C. A. Colla, A. Racoveanu, K. R. Coffee, A. F. Panasci-Nott, E. M. Kahl and J. G. Reynolds, *J. Phys. Chem. C*, 2022, **126**, 18802–18812.
4 J. G. Reynolds, A. Racoveanu, A. K. Burnham, J. S. Moore, K. R. Coffee, A. F. Panasci-Nott, K. D. Morrison, B. Koroglu, G. L. Klunder, C. A. Colla, J. Lee, H. E. Mason, J. D. Van Horn and E. M. Kahl, *APS SCCM Meeting*, Chicago, IL, 19–23 June, 2023, LLNL-PRES-850235.
5 P. C. Hsu, G. Hust, W. M. Howard and J. L. Maienschein, *Proceedings of the 14th International Detonation Symposium*, 2010, pp. 984–990.
6 J. G. Reynolds, N. K. Muetterties, A. J. Nelson, H. E. Mason, J. S. Moore, K. R. Coffee and E. M. Kahl, *Propellants, Explos., Pyrotech.*, 2021, **46**, 1–15.
7 J. F. Wardell and J. L. Maienschein, *Proceedings of the 12th International Detonation Symposium*, Office of Naval Research, ONR-333-05-2, San Diego, CA, 2002, pp. 384–393.
8 J. S. Moore, M. A. McClelland, P. C. Hsu, G. F. Ellsworth, E. M. Kahl and H. K. Springer, *Proceedings of the 16th International Detonation Symposium*, Cambridge, MD, 2018, pp. 936–945.
9 J. S. Moore, K. D. Morrison, A. K. Burnham, A. Racoveanu, J. G. Reynolds, B. Koroglu, K. R. Coffee and G. L. Klunder, *Propellants, Explos., Pyrotech.*, 2023, **48**, e202300237.
10 K. R. Coffee, J. S. Moore, A. Racoveanu, K. D. Morrison, A. K. Burnham, B. Koroglu, G. L. Klunder and J. G. Reynolds, *TATB thermal decomposition: An improved kinetic model for explosive safety analysis, International Explosives Conference*, Edinburgh, Scotland, 18–20 June, 2024.
11 E. G. Prout and F. C. Tompkins, *Trans. Faraday Soc.*, 1944, **40**, 488–498.
12 R. R. McGuire and C. M. Tarver, *Proceedings of the 7th International Detonation Symposium*, Naval Surface Weapons Center NSWC MP 82-334, Annapolis, MD, June 10–19, 1981, pp. 56–64.
13 C. M. Tarver, S. K. Chidester and A. L. Nichols, *J. Phys. Chem.*, 1996, **100**, 5794–5799.
14 M. L. Hobbs and M. J. Kaneshige, *J. Chem. Phys.*, 2014, **140**, 124203.
15 A. K. Burnham, K. R. Coffee, G. L. Klunder, A. F. Panasci-Nott and J. G. Reynolds, LLNL report JRNL-848399, *Propellants, Explos., Pyrotech.*, 2023, **48**, e202300121.
16 T. A. Land, W. J. Siekhaus, M. F. Foltz and R. Behrens Jr., *Proceedings of the 10th Symposium (International) on Detonation*, Boston, MA, 1983, pp. 181–189.
17 B. J. Yancey, N. K. Muetterties, E. M. Kahl, E. A. Glascoe and J. G. Reynolds, *Proceedings of the 16th Symposium (Inter-national) on Detonation*, ONR-43-5762-19, 2018, pp. 1664–1675.
18 K. D. Morrison, A. Racoveanu, J. S. Moore, A. K. Burnham, B. Koroglu, K. R. Coffee, A. F. Panasci-Nott, G. L. Klunder, B. A. Steele, M. A. McClelland and J. G. Reynolds, LLNL report, JRNL-852686, *Sci. Rep.*, 2023, **13**, 21256.

19 K. D. Morrison, J. S. Moore, K. R. Coffee, B. Koroglu, A. K. Burnham and J. G. Reynolds, LLNL Report JRNL-853782, *Propellants, Explos., Pyrotech.*, 2023, **48**, e202300268.

20 A. K. Burnham, K. R. Coffee, G. L. Klunder, A. F. Panasci-Nott and J. G. Reynolds, LLNL report JRNL-848399, *Propellants, Explos., Pyrotech.*, 2023, **48**, e202300121.

21 K. R. Coffee, A. F. Panasci-Nott, J. A. Olivas, J. Selinsky, K. D. Morrison, A. K. Burnham, G. L. Klunder and J. G. Reynolds, Analysis of degradation products in thermally treated TATB, *Propellants, Explos., Pyrotech.*, 2024, **49**, e202300176.

22 K. R. Coffee, A. F. Panasci-Nott, B. J. Stewart, J. A. Olivas, A. M. Williams and J. G. Reynolds, *Propellants, Explos., Pyrotech.*, 2022, **47**, e202100224.

23 J. Sharma, W. L. Garrett, F. J. Owens and V. L. Vogel, *J. Phys. Chem.*, 1982, **86**, 1657–1661.

24 R. Belmas, A. Bry, C. David, L. Gautier, A. Keromnes, D. Poul-Lain, G. Thevenot, C. Le Gallic, J. Chenault and G. Guillaumet, *Propellants, Explos., Pyrotech.*, 2004, **29**, 282–286.

25 J. S. Moore, K. D. Morrison, A. K. Burnham, A. Racoveanu, B. Koroglu, K. R. Coffee, A. F. Panasci-Nott, G. L. Klunder, B. A. Steele, M. A. McClelland, J. D. Van Horn and J. G. Reynolds, Lawrence Livermore National Laboratory Report, LLNL-TR-850231, DOPSR 23-S-2313, 2023.

26 J. S. Moore, A. K. Burnham, K. D. Morrison, A. Racoveanu, K. R. Coffee, B. Koroglu, G. L. Klunder and J. G. Reynolds, *APS Shock Compression of Condensed Matter Conference*, Chicago, IL 6.20.23, Lawrence Livermore National Laborastory Report, LLNL-PRES-849784, 2023.

Using AI for the discovery and synthesis of advanced energetic materials

Sandeep S. Tamber,*[a] Claire J. Taylor,[b] Declan H. Williams,[c] Javid Hamid,[c] Alison Beardah,[d] Justin Fellows,[d] Hugh Patrick,[d] Elizabeth Swann,[d] William Payne[d] and Charles J. M. Footer[a]

During the past few decades, enormous resources have been invested in research efforts towards the discovery of new energetic materials with improved performance over those in current use. The development of novel materials requires an extensive and time-consuming screening methodology to determine viable routes to manufacture. This is further compounded by requirements for the use of expensive quantum chemistry calculations requiring significant computational power. This activity is also heavily reliant on the use of energetic materials subject matter experts (SMEs) to oversee the entire process. This research has explored the development of a centralised Advanced Energetic Materials (AEM) System combining a suite of AI tools for the generation of novel energetic molecules, prediction of their properties and synthesis routes, theoretical validation of energetic performance and pathway viability. The natural language processing (NLP) and machine learning research areas within the artificial intelligence (AI) space have become more sophisticated with models able to carry out a range of activities such as question-answering (Q&A) with greater ease and efficiency. This paper describes the development of a large language model (LLM) system to query an energetics-based repository of data that is used to underpin the AEM system.

1 Introduction

The research described in this paper aims to accelerate the discovery and delivery of novel explosives and energetic materials. As a critical, underpinning technology, new energetic materials are required for a wide range of defence and security applications.

With ever-increasing demands on the performance of next generation weapons systems there is a requirement for development of novel energetic materials to fulfil this role.

The Advanced Energetic Materials (AEM) platform developed has significant potential to realise important benefits – including reductions in both manufacturing costs and environmental impact as well as improved energetic performance. The aim is to harness new and under-exploited technologies, with a focus on automated processes, to identify new energetic materials, as well as discovery of new synthetic pathways for these materials.

[a] *QinetiQ, Cody Technology Park, Ively Road, Farnborough, GU14 0LS, UK. E-mail: sstamber@qinetiq.com*
[b] *QinetiQ, Malvern Technology Centre, St Andrews Road, Malvern, Worcestershire, WR14 3PP, UK*
[c] *QinetiQ, Fort Halstead, Sevenoaks, Kent, TN14 7BP, UK*
[d] *DTSL, Porton Down, Salisbury, UK*

At the heart of the AEM platform is a Question and Answer (Q&A) system that can be used by materials scientists to interrogate the energetics-based repository of data that is used to underpin the AEM. Creating an effective Q&A system for specialised domains, such as AEMs, presents unique challenges and opportunities due to the highly specialised and technical nature of the field. AEMs, which include substances that are capable of rapid energy release, are critical in various forms of applications, ranging from industrial processes to defence. This research paper focuses on the development of a large language model (LLM) approach to a Q&A system specifically designed for a repository of knowledge on AEMs.

The research has leveraged cutting edge NLP techniques and domain specific expertise. The system provides precise and contextually accurate responses to complex queries, thereby enhancing access to, and understanding of, this intricate field for not only SMEs but lay audiences.

During this work, two Q&A systems were developed. The first was an AEM specific BERT LLM and the second used the Mistral LLM within a retrieval augmented generation pipeline with FAISS retrieval technology.[2,10] Both techniques and their subsequent findings are discussed in Sections 1 and 2 of this paper. By developing two distinct models, we were able to exploit the well-understood deep contextual understanding of the BERT and compare it against innovative retrieval augmented generation (RAG) technologies. The integration of these techniques significantly enhances the retrieval and application of knowledge in this specialised domain.

1.1 Method

The research was scoped to integrate machine learning (ML) with existing AEM synthesis processes, productivity tools and feature prediction algorithms on a local air-gapped (no network connections) server. The server contains open source research papers, and user interfaces for SME's to interact with the system. Server specifications relevant to the research will be described within the results section of the paper. All technologies used in this project are open source apart from the Quaisr platform used to host the QinetiQ code.

The repository of data consisted of open source scientific papers and these were collected from two publication bodies: ArXiv and the Institute of Electrical and Electronics Engineers (IEEE) in accordance with each of the repositories terms and conditions. There was an expectation that peer reviewed scientific articles would consist of high quality text, be devoid of colloquial language and have consistent formatting for data pre-processing.[1] Table 1 below summarises search terms related to energetic materials across the three repositories along with their paper counts. The final research paper count in the platform database was 271.

The collected data was integrated into the training AEM LLMs for the AI stream and for the natural language to provide the LLMs with context. The former approach was used for open domain Q&A development and the latter for closed domain Q&A respectively.

Table 1 Search term paper counts across different data repositories (accessed on 14/04/2024).

Search term	ArXiv	IEEE
High energy materials	6	37
high energy density	638	2829
energetic materials	48	146
explosives AND energetics	0	116
Propellants AND energetics	0	60
High energy density materials	23	6
primary explosives	5	2
Polycyclic energetic materials	0	0
Total research paper count	722	3196

1.1.1 AEM BERT development. Open domain Q&A requires three critical components: an input question provided by the user, a LLM, and a context string, which is a piece of text that holds the answer to the question. The LLM technologies used within this pipeline were variations of Google's Bidirectional Encoder Representations Transformer, BERT.[2,3] There have been notable successes using BERT to create domain specific LLMs, such as BatteryBERT, SciBERT and ChemBERTa.[4–6] BERT training entails tokenising input data and embedding into vectors before processing into the model. The BERT then undergoes multiple layers of bidirectional transformations to capture contextual relationships. Training can either be from first principle to create a LLM from only the training data or using a method known as transfer learning to imbue the additional AEM research into an existing trained base BERT LLM.

There are many variations of the BERT LLM that can be used as a base model. The most popular and widely baseline model for training is bert-base which was trained on BooksCorpus and the English Wikipedia.[2] However, data scientists can also opt to use BERT variants that have already had transfer learning conducted on them such as SciBERT or MatSciBERT. The advantage of using transfer learned models is that users can build upon the knowledge already imprinted into LLMs with their own data, and this requires significantly less data when compared to training a foundation LLM.

Although, comparatively smaller amounts of training data are required for transfer learning, the quantity of papers required is still substantial. Within research, it has been observed that large collections of data are required to pre-train a model so that is able to perform sufficiently well for downstream tasks such as Q&A. For instance, BatteryBERT is a domain specific LLM used for battery enhancement research.[7] The BatteryBERT researchers collected 487 484 scientific papers for the pre-training and other down streaming tasks.[4] Similarly, another top performing BERT model SciBERT used 1.4 million papers for its training.[5]

Generating a domain specific BERT is a two-step process, consisting of pre-training and fine-tuning. Pre-training involves two separate tasks: masked language modelling (MLM) and next sentence prediction (NSP). In MLM, researchers hide words in the corpus of training data with a token, ⟨MASK⟩, which is relevant to the domain of knowledge and also not included in a list of stop words (the, and, they, he, she *etc.*). Once the masking is complete the BERT model is asked to predict what the masked word should be. See Fig. 1 below for an example of masking.

NSP is the second step of the pre-training process. A number of different corpora of research papers are split by sentences. The remaining sentences are placed into a labelled dataset of sentence pairs where either one sentence follows another or two completely different sentences are joined together. The data is supplemented with a binary label column, 1 if the sentence follows another in the corpus or 0 if they are unrelated. The BERT model is then used to determine which binary label to assign to each sentence pair. The attempts of the BERT model to predict the masked words and predict sentence pairs provide the weights to create a domain specific LLM. There are many methods to determine the effectiveness of a BERT model. Our research used a simple MLM task to determine how well the model performed when predicting words in an unseen dataset after pre-training was complete. See Fig. 2 for the complete BERT training process.

The pre-trained model is then fine-tuned to initialise a layer for the downstream Q&A task. The BERT model was provided with labelled question-answer pairs from the Stanford Question Answering Dataset (SQuAD) and asked to predict the answers to each question. SQuAD is a reading comprehension dataset compiled from more than 100 000 questions from Wikipedia articles.[8] Exact match metrics measure the percentage of predictions that match the ground truth answers exactly. The end product of this training stream was an AEM BERT tailored

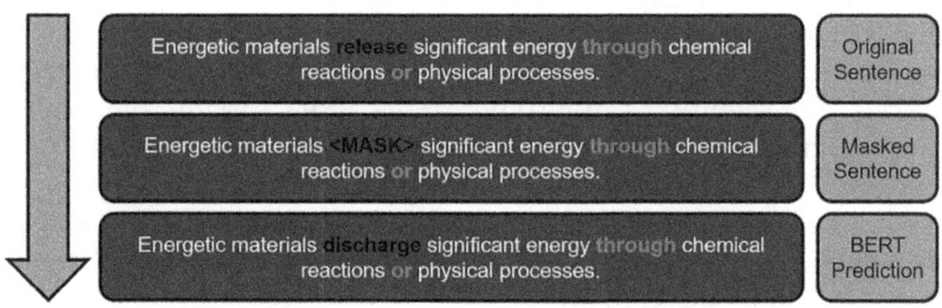

Fig. 1 MLM process for BERT pretraining. Words in blue are stop words and not included in the masking process. The words in black have been masked in the second step of the process and the BERT predicts what it could be in the last step. The output is fed back into the training loop.

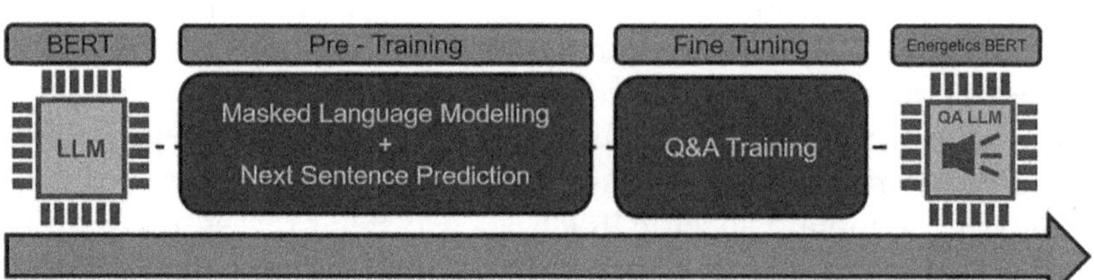

Fig. 2 Process for applying transfer learning to an existing BERT model. The BERT is pre-trained and fine-tuned to produce an AEM Q&A version of the original model.

for Q&A. Through an integrated UI system, researchers and SME's can quickly re-train or create new domain specific models with little technical knowledge, test the output models, change the training parameters, update the downstream tasks for the pre-trained models, update the foundation models for pre-training and integrate Q&A across other ML streams within the system.

1.1.2 AEM RAG development. Unlike open domain Q&A as described above, closed domain Q&A requires an LLM and a query. Unlike open domain Q&A, the context can be optionally supplied to supplement the LLMs answer. Generative LLMs were used for this phase of research. Generative LLMs are AI models that are able to generate contextually appropriate and coherent text by leveraging extensive training datasets to predict and produce human like text. The training process for embedding a specific domain of knowledge into a LLM is also known as pre-training, although the techniques to do this may vary in comparison to the BERT pre-training.

Beltagy *et al.*[5] and Lews *et al.*[9] suggested that pre-training a tokeniser might improve the subsequent pre-training of the BERT; a script was developed to do this however very little improvement was observed. Authors suggest that was likely due to the low volume of data sourced within the data repository at the point of experimentation.

For the closed domain Q&A, authors have developed a methodology known as "basic" RAG.[10] The RAG workflow requires a repository of one or more input text, a question and a LLM, usually a generative type. Initially, the text in the repository of information is converted into numerical values and indexed into a vector database using a separate embedding LLM. We opted to use sentence-transformers/multi-qa-MiniLM-L6-cos-v1, as it held the best benchmarks at the time of the experiment. The question is also passed through an embedding LLM and the output vector then used to determine which values in the vector database are semantically most similar to the query. This process is used to filter out relevant text from the vector database. Retrieval filtering not only allows users to understand the providence of the information source but also reduces the text comprehension workload on the LLM. The query and retrieved text are then combined into a prompt, which is parsed into the LLM to generate an answer, see Fig. 3 below. The method will be further explored in the results section of this paper.

Meta's Facebook AI Semantic Search (FAISS) is the retriever technology within the RAG pipeline to extract the relevant data from a vector database.[11,12] The FAISS toolkit manages the indexing methods, searching, core functionality of the vector databases as well as acting as a database management system. Once the data has been embedded into vector space, users can compute a search between the query vector and the embedded data.

The FAISS library is able to embed the data into a number of different architectures, known as indexes: Flat, Hierarchal Navigable Small Word (HNSW), Local Sensitive Hashing (LSH), Inverted File Index product quantisation (IVFPQ) along with hybrid architecture variants.[12] All of the different methods to index the data were explored within this study. The

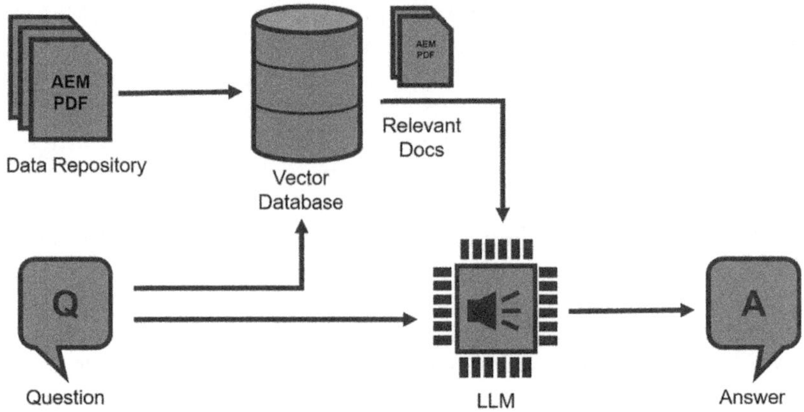

Fig. 3 Simple RAG process used for closed domain Q&A. PDFs are embedded into a vector database, relevant documents are retrieved according to the input query and both are finally ingested into a LLM to generate an answer.

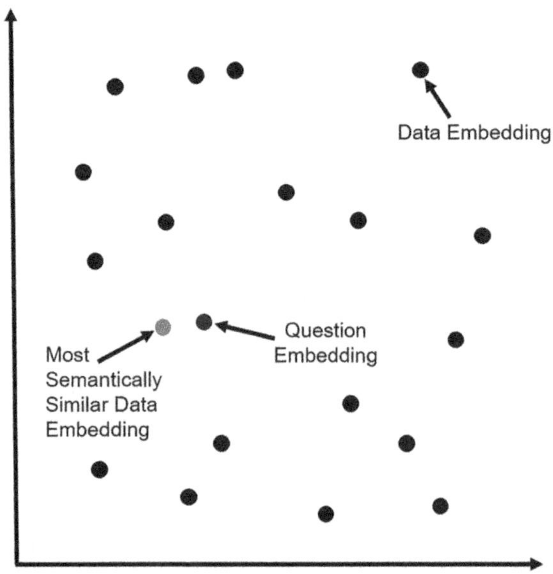

Fig. 4 Two dimensional representation of the FAISS data, research paper, embedding's into vector space. The Euclidian distance between the question embedding and the data points was calculated to find the data embedding with the closest semantic similarity. This in theory should hold the answer to the question asked by the user.

indexing methods all had a trade-off between the speed of the retrieval and the quality of the index. However, due to the high compute power available and the low number of papers required to be embedded, the indexing method with the highest quality, the Flat Index, was chosen as the benchmark for document retrieval. Flat indexing computes the Euclidian distance between the data, research paper, vectors and the question embedding, retrieving the k articles closely aligned to the question semantically. The variable k was an integer value that determined the number of articles output by the retrieval process. See Fig. 4 below.

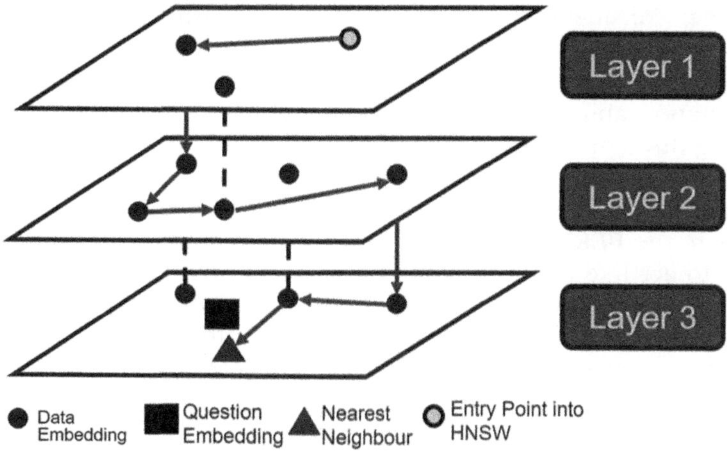

Fig. 5 HNSW architecture. The data embedding are distributed across many layers in this architecture with some data embedding overlapping. The query embedding is projected onto one of these layers and the algorithm works to reach it from an entry point. Once the algorithm reaches the question embedding layer, a Euclidian distance is calculated to find the data embedding closest to the question.

If the following passage is relevant, answer this question: **[QUESTION]**. Otherwise still attempt to answer the question with the given data but caveat it may be incorrect: **[RETRIEVAL PASSAGE]**.

Fig. 6 Example of a LLM prompt used for Q&A within the workflow. The [QUESTION] and [RETRIEVAL PASSAGE] variables were updated according to the user question and the FAISS retrieved passage.

If retrieval performance ever becomes an issue in the future, a HNSW architecture was also developed. In this instance, the data points were embedded onto a number of different layers connected *via* links.[12] The data is then retrieved through this architecture using connections between the layers. The relevance of the retrievals brought back may be of poorer quality, but retrieval speed is increased (Fig. 5).[12]

Once relevant retrievals are identified, they are merged into a prompt, which is passed to a LLM. This is known as prompt engineering. Prompt engineering is a critical methodology in harnessing the capabilities of a LLM. This involves developing precise prompts tailored to the task at hand enabling LLMs to generate coherent, informative and contextually relevant outputs. The prompt engineering process requires careful tuning to ensure that the restraints on it are not too tight so as to restrict the LLM from augmenting the retrieval with its own knowledge, but also not too loose as to cause the LLM to hallucinate. An example of a prompt used in the research work can be seen below in Fig. 6.

Selecting the appropriate LLM for RAG is a multifaceted process that requires careful consideration of many factors. Ideally, the chosen LLM should display strong generative capabilities to produce high quality outputs. The incorporation of evaluating LLM benchmarks against established metrics is a useful method to understand its capabilities. Comparing LLMs against benchmarks also allows researchers to identify models that excel in the generation of informative and coherent responses.

Another consideration for this work was ensuring that the models could be hosted locally on a server. The Ollama software was therefore chosen to run open-source LLMs locally. Ollama bundles model weights, configuration and data into a single model file that can be called for querying through an interactive shell, REST API or a python library. For our experiments the Mistral LLM was used.[13]

The RAG's retrieval capability was also leveraged to provide context strings to the BERT Q&A model. This hybridisation allowed the BERT model to act like a closed domain Q&A, and allowed users to simultaneously run a number of LLMs with different architectures. This novel technique is useful for the ML community as it will allow data scientists to use previously generated BERT LLMs with newer industry technologies.

2 Results and discussion

2.1 AEM BERT results and findings

Conducting transfer learning on LLMs was an energy and cost intensive process. The bottlenecks for this process were the computing power available on the server and the number of research papers to process. The server held multiple GPUs (NVIDIA Tesla A100 Ampere 40GB). When the GPUs were not initialised the BERT training times were ~3 hours but when activated, training was completed within ~5 minutes. The huge reduction in training time indicates the power of leveraging hardware when training AI. The low training times also likely contributed to the significantly small training dataset being parsed into the BERT model. A batch size of 64 and a learning rate of 1E-5 over 64 epochs was found to be the optimum set of hyper-parameters for the BERT pre-training.

BERT pre-training was conducted on bert-base and SciBERT LLMs with the same experimental hyper parameters. Both output models showed an increase in MLM performance after the transfer learning was complete It was noted that the SciBERT outperformed BERT by 12%, this is likely due to the same linguistic similarities of the base model as the peer reviewed articles used to train our domain specific AEM model. See Table 2 below for the experimental MLM results.

Although it can clearly be seen that the pre-training increases the performance of the BERT model, it is still far below what we would expect from a LLM for a productionised system. The speculated reason for this shortfall was the small repository of available data. In the methodology

Table 2 MLM accuracy metrics for BERT models before and after AEM transfer learning.

BERT LLM	MLM accuracy (%)
BERT-BASE	18
SciBERT	30
BERT-BASE after pre-training	37
SciBERT after pre-training	42

section, previous research has shown that it could be beneficial to train a tokeniser when attempting to train a domain specific LLM. However, the experiments showed no difference in performance with the untrained datasets. Although the BERT LLMs were fine-tuned to enable Q&A capabilities, the answers provided by the models were generally not fluent, took a long time to execute and were largely incorrect. Close domain Q&A was pursued thereafter due to integration of cutting-edge techniques available with generative foundation models and retrieval technologies. However, the workflow to create domain specific BERT models will be relevant in the future as it is expected that the repository of data will increase over time.

2.2 AEM RAG results and findings

Working within a RAG framework had a number of benefits. Firstly, since the LLM was primarily used as a comprehensive technology, there is a lot more freedom in deciding which foundational LLM to use. Currently there are constant developments taking place in the NLP and generative space with numerous models published on a regular basis. The RAG process allows users to simply change the LLM without impacting the rest of the workflow architecture.

Secondly, the knowledge embedded into a LLM was strictly restricted to its training knowledge, *i.e.* if there is a new development in energetic materials, the LLM trained on legacy data would not necessarily have any knowledge about it. Therefore, having a retrieval step in the Q&A process ensures that as long as the relevant data exists in the repository, researchers would always be able to extract the knowledge with Q&A. Thirdly, the entire process is data agnostic. If the user decided they would like to specialise within a field of AEM all they need to do is update the repository of information. Alternatively, the user could also completely change all the data and the workflow, and it would still operate.

Finally, as previously mentioned the domain of NLP is experiencing a vast level of innovation and growth. There are numerous technologies in development. Due to the modular nature of RAG, entire sections (LLMS, retrieval technology, prompts) were updated with better tools with no disruption to the architecture required for the overall workflow to function. The powerful modular nature of this workflow allowed technologies to be consistently upgraded and reduced the issue of product redundancy.

During the course of this work, research in the AI community has speculated that the larger batch sizes for the FAISS indexing adversely affects the retrieval of relevant text for a given query.[14] To test this theory a collection of Q&A pairs were manually derived from the abstracts of 90 scientific papers. A number of different Flat Indexes were generated using various batch sizes (4, 8, 12, 32, 64, 128) and then the five top retrievals were extracted for the given queries. There was found to be 0 variance between the outputs, indicating that batch sizes do not affect the retrieval ability in the FAISS system.

A pre-processing parameter considered when building the FAISS index was the chunking value. The chunking value determines the character

Table 3 Investigation of different batch sizes and chunk sizes on the FAISS retrieval ability.

Hyperparameters		
Batch size	Chunk size	Percentage accuracy (%)
4	250	43
4	500	47
128	250	43
128	500	46

length split of the ingested research articles *i.e.* if the paper is 30 000 strings long, and the chunking value is 1000 the text will be split into 30 equal parts. Chunking the text for indexing is important as lengthy strings will result in slow retrieval processes with the context sizes overwhelming the LLM. Conversely, if the chunk length is too small the retrieved text may lack nuance. To determine the optimum text length, experiments were carried out where batch and chunk size were varied for the same input data as above. As can be seen from the Table 3 below a batch size of 4 with a 500 chunk size had the highest retrieval accuracy.

To understand the retrieval ability of FAISS, 50 questions were generated from the data repository and the retriever was asked to recall the top five ($k = 5$) retrievals. An accuracy of 68% was observed in this experiment. The retrieved strings have a pivotal importance in the performance of the RAG workflow, especially if the prompt provided within the workflow restricts LLM answer augmentation. There are a number of methods to improve performance such as chunk size optimisation (as done above), sliding chunk windows, metadata attachments, query routing, re-ranking results *etc.*

Re-ranking the results is an optimisation technique built into the FAISS technology which was initialised for the workflows. Re-ranking ingests the top retrievals from FAISS and adjusts their order of relevance. The technique was especially useful to implement with the FAISS technique as it usually retrieves the correct documents over $k = 5$ retrievals, but the top initial retrieval ($k = 1$) is not necessarily the most relevant to the question, which may for instance be in $k = 3$ position. Re-ranking attempts to rectify the order mismatch within the retrieved texts.

The user interface was hosted on a third party app by Quaisr.[15] After various rounds of SME testing, developers aimed to mimic typical chatbot type interfaces as closely as possible, in order to develop an intuitive and easy to use platform. Users can ask the interface a question and are able to audit their answer with the ability to see the context retrieved and also download the content file from the repository of data it was extracted from. See Fig. 7 below.

The ability to understand the providence of the source data is key for users and typically not available in most chatbot type LLMs which have no audit trail. We have successfully ensured that this system is not a black box for users, and is of paramount importance for fields such as defence and engineering.

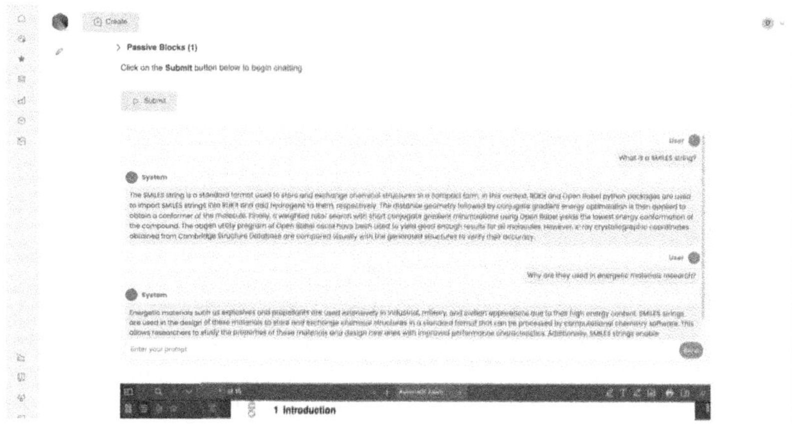

Fig. 7 Screenshot of SME chat bot user interface for the RAG workflow on the Quaisr platform developed by authors. Retrieved research papers by the RAG are viewable and downloadable at the bottom of the page.

In future iterations of the project more robust evaluation processes will be developed, however, the outputs of the RAG system were evaluated by material science SME's who reported the following eight insights:

1. RAG generated better answers for technical questions
2. RAG gave more specific answers for specific questions
3. RAG regularly bolstered the given context with its own knowledge
4. Chatbot knowledge is limited to the date it was created so it was not useful for cutting edge research
5. On occasions the LLM refused to output answers if they were related to energetics
6. Speed of the RAG BERT was the slowest, an out-of-the-box model had the fastest performance
7. RAG tends to hallucinate less than an out-of-the-box LLM
8. RAG is an explainable AI solution as it references the original data

3 Conclusions

In conclusion, we have successfully developed a user friendly, explainable AI solution which also produces relevant results for AEM researchers in a completely air gapped and secure space for sensitive corporate and government data. The AEM tailored RAG workflow shows the most promise as a tool for the advancement of AEM, streamlining the process of information retrieval, and future proofing Q&A capabilities with its modular architecture. Additionally, we introduced an AEM BERT designed for Q&A. Although the performance currently needs to be improved, we have developed functioning workflows and once more data can be sourced the BERT can be quickly retrained. These advancements mark significant progress in leveraging AI for AEM research. Looking forward, we anticipate further refinement of tools within the Q&A processes, expansion to innovative AI technologies and inclusion of further research paper repositories in order to broaden their applicability and impact in this field.

Acknowledgements

The authors would like to thank the teams at QinetiQ with their support for this project, the developers at Quaisr for their support with their platform, and finally Dstl for their permission to publish this paper and funding for the project.

References

1 J. Hoffmann, S. Borgeaud, A. Mensch, E. Buchatskaya, T. Cai, E. Rutherford, D. de Las Casas, L. A. Hendricks, J. Welbl, A. Clark, T. Hennigan, E. Noland, K. Millican, G. van den Driessche, B. Damoc, A. Guy, S. Osindero, K. Simonyan, E. Elsen, J. W. Rae, O. Vinyals and L. Sifre, Training Compute-Optimal Large Language Models, *arXiv*, preprint, 2022, arXiv:2203.15556.

2 J. Devlin, M. W. Chang, K. Lee and K. Toutanova, BERT: Pre-training of Deep Bidirectional Transformers for Language Understanding, *arXiv*, preprint, 2018, arXiv:1810.04805.

3 A. Vaswani, N. Shazeer, N. Parmar, J. Uszkoreit, L. Jones, A. N. Gomez, Ł Kaiser and I. Polosukhin, Attention is All You Need, in *Advances in Neural Information Processing Systems*, 2017, vol. 30, pp. 5998–6008.

4 Z. Huang, Y. Wen, Z. Jia, W. Wang, L. Liu, D. Zhang, Y. Zhang, H. Chen and L. Zhang, BatteryBERT: A Pretrained Language Model for Battery Database Enhancement, *arXiv*, preprint, 2023, arXiv:2306.10856.

5 I. Beltagy, K. Lo and A. Cohan, SciBERT: A Pretrained Language Model for Scientific Text, *arXiv*, preprint, 2019, arXiv:1903.10676.

6 S. Chithrananda, G. Grand and B. Ramsundar, ChemBERTa: Large-Scale Self-Supervised Pretraining for Molecular Property Prediction, *arXiv*, preprint, 2020, arXiv:2010.09885.

7 Hugging Face, What is Question Answering? https://huggingface.co/tasks/question-answering (accessed Jan 14, 2024).

8 P. Rajpurkar, J. Zhang, K. Lopyrev and P. Liang, SQuAD: 100 000+ Questions for Machine Comprehension of Text, *arXiv*, preprint, 2016, arXiv:1606.05250.

9 P. Lewis, E. Perez, A. Piktus, F. Petroni, V. Karpukhin, N. Goyal, H. Küttler, M. Lewis, W. T. Yih, T. Rocktäschel, S. Riedel, D. Kiela, Retrieval-Augmented Generation for Knowledge-Intensive NLP Tasks, in *Advances in Neural Information Processing Systems*, 2020, vol. 33, pp. 9459–9474.

10 Facebook Engineering, Faiss: A Library for Efficient Similarity Search. https://engineering.fb.com/2017/03/29/data-infrastructure/faiss-a-library-for-efficient-similarity-search/ (accessed Feb 7, 2024).

11 J. Johnson, M. Douze and H. Jégou, Billion-Scale Similarity Search with GPUs, *IEEE Trans. Big Data*, 2019, **5**(3), 557–567.

12 J. Briggs, Faiss: The Missing Manual, Pinecone. https://www.pinecone.io/learn/series/faiss/ (accessed Feb 7, 2024).

13 Ollama, Ollama. https://ollama.com/ (accessed Feb 16, 2024).

14 K. Elena, Why You Should Be Careful Using FAISS, Medium. https://medium.com/@elenek/why-you-should-be-careful-using-faiss-c44996eda9ee (accessed April 19, 2024).

15 Quaisr, Quaisr. https://www.quaisr.com/ (accessed Aug 9, 2024).

Numerical framework for the optimization of HVRB parameters from experimental measurements

Simone Falco,*[a] Liam Smith,[a] David Chapman,[a] Ajen Limbu,[b] Mark Whittaker,[b] Daniel Pope,[b] Nik Petrinic[a] and Daniel Eakins[a]

Modelling energetic materials is commonly performed in an Eulerian framework, inherently suitable for simulating fluid-like behaviour of reaction products and the propagation of pressure waves in surrounding fluid media (*e.g.* air or water). Conversely, when seeking to simulate the mechanical response of unreacted material, a Lagrangian framework is instead preferable. This difference raises a question of how best to simulate the thermo-mechanical response of damaged explosives. In this paper a numerical optimisation framework is detailed, parameterising an in-house Lagrangian implementation of the History Variable Reactive Burn (HVRB) model from measurable experimental results. To ensure the computational efficiency and stability of the optimisation process, a fully automated sensitivity study is performed across the parameter set, grouping variables into relevant concurrent optimisation processes. The numerical optimisation of the full set of HVRB parameters utilises the commercial software LS-Opt, combining on-the-fly generation of Finite Element (FE) models, subsequent simulation of mechanical insult driven detonations, and post-processing of data mimicking experimental procedures. This workflow provides the necessary tools for the relatively rapid assessment of explosive materials, allowing the prediction of complex behaviour from limited experimental datasets. Ultimately this paves the way to linking changes in observed detonation behaviour (*e.g.* due to mechanical damage) to underlying parameter variabilities.

1 Introduction

1.1 Background and motivations

The modelling of energetic materials relies on the accurate evaluation of the numerical parameters for the equation of state (EoS) used to describe their behaviour. Whilst some parameters are "physically-based", and can therefore be extrapolated using physical considerations, most are generally empirical parameters that need to be reverse-engineered by fitting the numerical results to experimentally measured curves.

The (HVRB) model,[1] which has proven to be very effective to model the initiation and propagation of the reaction in energetic materials using a relatively small number of parameters, is an example of a constitutive model requiring a set of parameters that are not directly linked to any specific material properties.

Building on the implementation of HVRB in the Lagrangian code LS-Dyna by the authors,[2] the work presented in this paper details the implementation of an automated approach to extract the relevant

[a] *University of Oxford, UK. E-mail: simone.falco@eng.ox.ac.uk*
[b] *Dstl, Porton Down, UK*

constitutive parameters from an arbitrary set of experimentally measurable data representing the material response to specific loading conditions.

The approach, detailed in Section 2, relies on two algorithms. The first evaluates the correlation between the data available and the parameters, identifying which parameters can be evaluated, and their reasonable range, while the second carries out the actual evaluation of the optimal value for each parameter.

1.2 HVRB formulation

HVRB is a constitutive model originally implemented in the Eulerian code CTH[3] which is used for the Finite Element simulation of energetic materials. The HVRB formulation introduces the extent of reaction (λ) to control the reaction, effectively defining a mixture rule for pressure (P) and energy (E) that describes the transition from reactant (P_r, E_r) to product (P_p, E_p) EoS, as reported in eqn (1).

$$\begin{cases} E(\rho, T, \lambda) = (1-\lambda)E_r(\rho, T) + \lambda E_p(\rho, T) \\ P(\rho, T, \lambda) = (1-\lambda)P_r(\rho, T) + \lambda P_p(\rho, T) \end{cases} \tag{1}$$

where ρ and T are the density and the temperature, respectively, assumed uniform within the element.

The extent of reaction (λ) can be calculated from the value of pressure (P) as:

$$\lambda(t) = \min\left(1, \phi(t)^M\right) \tag{2}$$

$$\phi(t) = \frac{1}{\tau_0}\int_0^t \left[\frac{(P - P_I)}{P_R}\right]^Z \mathrm{d}t \tag{3}$$

where τ_0, P_I, P_R, M, and Z are material parameters.

Whilst τ_0 is generally considered invariant for all materials, and in this paper imposed equal to 1 ms, the other four parameters are strongly material dependent, and their optimisation is the main objective of the work presented in this paper.

HVRB was implemented in the Lagrangian code LS-Dyna[4] as a "user defined equation of state" by the authors, enabling the use of Lagrangian contact formulations, which are notably more robust and computationally efficient than the Eulerian counterparts. In this paper the Lagrangian formulation has been used to simulate shock-to-detonation transition (SDT) test consisting of a solid projectile impacting an energetic material specimen at a user-defined velocity, generating a uniaxial shock wave within the explosive that in turn triggers the detonation at a distance from the impact point. Owing to the potential for HVRB to reproduce pressure-triggered detonation, and the relative simplicity of modelling mechanical insults within a Lagrangian formulation, the SDT test has been chosen to validate the optimisation approach presented in this paper.

The full details of the HVRB formulation can be found in the report by Starkenberg[5] and the specifics of the Lagrangian implementation and its validation against the Eulerian formulation are detailed in the paper by the authors.[2]

2 Algorithm implementation and results

The SDT test, reproducing mechanical initiation of detonation, is used to demonstrate the ability of the methodology developed to reverse-engineer the values of the HVRB parameters for an energetic material with a specific response to mechanical insult.

The test consists of a steel projectile impacting a laterally confined specimen, leading to the formation of a one-dimensional shock wave traversing the energetic material sample, which eventually leads to the transition to detonation at a specific distance, commonly defined as *run to detonation distance* (x_d) from the impact point, as presented in Fig. 1.

To understand the relationship between HVRB parameters and the output of the SDT test a "*baseline*" set of parameters, not representative of any specific material, is assumed, as reported in Table 1. For an impact velocity of 700 m s^{-1}, the "*baseline*" model predicts a shock to detonation transition 5 mm away from the impact point (*i.e.* $x_d = 5$ mm), as shown by the velocity profiles in Fig. 1.

For the study presented in this work only the variability of the HVRB parameters is explored, whilst the behaviour of reactants and products, described by Grüneisen and JWL EoS, respectively, are assumed known with the relevant parameters taken from the open literature.[6]

The method implemented for the reverse-engineering of the parameters is based on the combination of two numerical processes: the sensitivity study and the parameter optimisation. The first evaluates the level of correlation between parameters and the desired output behaviours, as well as the meaningful range for each parameter under specific testing conditions. The second part of the strategy samples the domain defined by the range identified in the sensitivity study to find the

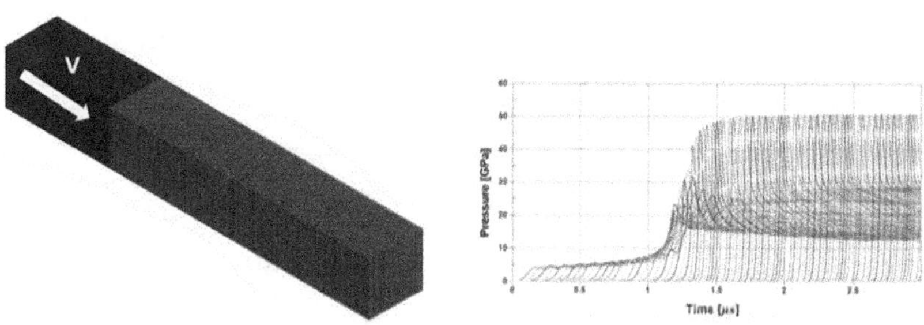

Fig. 1 Finite elements model of a shock to detonation (left) test consisting of steel projectile (dark) impacting laterally confined explosive specimen (light) at a user defined velocity and pressure evolution in each cross section (right) from the time of impact.

Table 1 Summary of the "baseline" HVRB parameters.

P_i	P_r	M	Z
0.05 [GPa]	1.0 [GPa]	3.5	1.0

optimal set of parameters that reproduce the specific material behaviours of interest.

The implementation of these two algorithms, and their use to extract the HVRB parameters to reproduce the "*baseline*" behaviour, are described herein.

2.1 Sensitivity study

To study the sensitivity of the simulation results to the HVRB parameters, variabilities of ±10%, ±20%, and ±50% are imposed to each of the four parameters individually, whilst holding the others equal to their baseline values as reported in Table 1. For each simulation three responses are measured: The run to detonation distance (x_d), the Chapman–Jouguet pressure (P_{CJ}), and the detonation velocity (v_d).

The combinations of 6 variations (±10%, ±20%, and ±50%) of four parameters (P_i, P_r, M, and Z) total 24 simulations, whose results, in terms of run to detonation distance, are summarised in Table 2 alongside the baseline simulation.

The central column shows the result of the baseline simulation predicting a run to detonation distance of $x_d = 5$ mm. The other cells represent the run to detonation distance (and the percentage difference to the baseline results) predicted by perturbing the parameter indicated in the row by the magnitude indicated in the column, whilst all the other parameters are kept equal to their baseline values. There are three important conclusions that can be drawn from the data summarised in Table 2.

The first is that the variation of P_i within the bounds explored is irrelevant on the prediction of the run to detonation distance as even a variation of ±50% leads to negligeable divergence from the baseline prediction.

The second conclusion is that the other three parameters are, conversely, all significantly affecting the overall behaviour, with P_r and M showing a variation of the result almost twice as large (in percentage terms) than the variation of the input, whilst even relatively small variation in the value of Z can lead to significantly different outcomes.

The last observation is that the variability chosen defines the parameter range of interest for this specific simulation, as different combinations of the three parameters within the range identified can reach both extreme behaviours of over-driven (OD) shocks (*i.e.* detonation occurring almost immediately upon impact) and significantly longer run to detonation distance which could be classified as non-detonation on

Table 2 Summary of the variability of the run to detonation distance for all the simulations carried out for the sensitivity study.

x_d [mm]	−50%	−20%	−10%	BL	10%	20%	50%
P_i	5.0 (0%)	5.0 (0%)	5.0 (0%)	5.0	5.0 (0%)	5.0 (0%)	5.0 (0%)
P_r	OD	2.6 (−48%)	3.6 (−28%)	5.0	6.8 (36%)	9.2 (84%)	20 (292%)
M	OD	3.0 (−40%)	4.0 (−20%)	5.0	6.2 (24%)	7.2 (44%)	10 (100%)
Z	>75.0	16 (228%)	8.8 (76%)	5.0	3.0 (−40%)	OD	OD

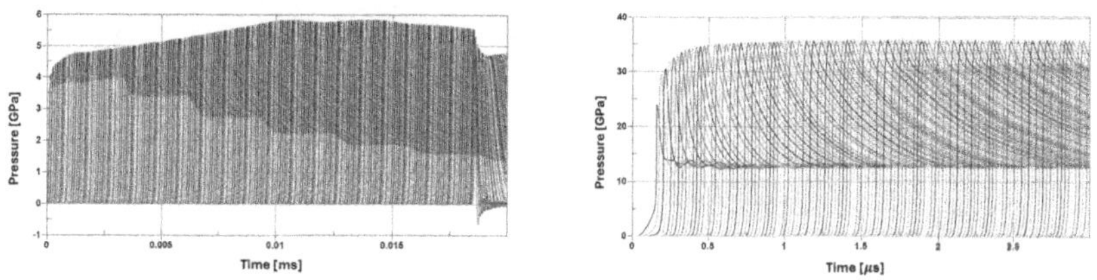

Fig. 2 Pressure profile of each cross section during an under-driven (left) and over-driven (right) test.

Table 3 Summary of the variability of the peak pressure for all the simulations carried out for the sensitivity study.

P_{CJ} [GPa]	−50%	−20%	−10%	BL	10%	20%	50%
P_i	50.7 (0%)	50.7 (0%)	50.7 (0%)	50.75	50.8 (0%)	50.8 (0%)	50.8 (0%)
P_r	OD	47.6 (−6%)	49.2 (−3%)	50.75	52.0 (3%)	53.0 (4%)	55.7 (10%)
M	OD	49.7 (−2%)	50.2 (−1%)	50.75	51.2 (1%)	51.5 (2%)	53.1 (5%)
Z	—	58.1 (15%)	54.8 (8%)	50.75	46.8 (−8%)	OD	OD

explosive specimens with regular length. The pressure profiles for the under-driven (*i.e.* not detonating) and over-driven cases are reported in Fig. 2.

The same approach is applied to evaluate the effect of the parameter variability on the peak pressure (P_{CJ}) showing a reduced level of correlation, as detailed in Table 3. Ignoring the cells leading to under- and over-driven cases, the discrepancy is generally <10%, with only one case of P_{CJ} being 15% greater than the baseline value, due to the variation of Z. Analysing the details of the data reported in the table, it is evident that the variation of the parameter P_i does not affect P_{CJ} either, whilst both P_r and M have a limited correlation, and Z has a small, but not negligible, effect.

The correlation between the detonation velocity and the HVRB parameters is significantly lower than observed for x_d and P_{CJ}, with the maximum discrepancy recorded being <1.5%, which is to be expected as the detonation velocity correlated mainly with the product EoS parameters. The detailed results are therefore not reported for concision.

The results of the sensitivity study can then be used to guide the setup of the optimisation study, ensuring the robustness and computational efficiency of the numerical process, and the accuracy of the results. Specifically, the relevant conclusions of the sensitivity study here are:

- SDT tests are not sensitive to the variation of the parameter P_i limiting the list of optimisable parameters to P_r, M, and Z;
- all three parameters have a similar range of interest, between −20% and 50% variation from baseline values;
- all three parameters correlate strongly with the run to detonation distance, mildly with the peak pressure, and show negligible correlation with the detonation velocity.

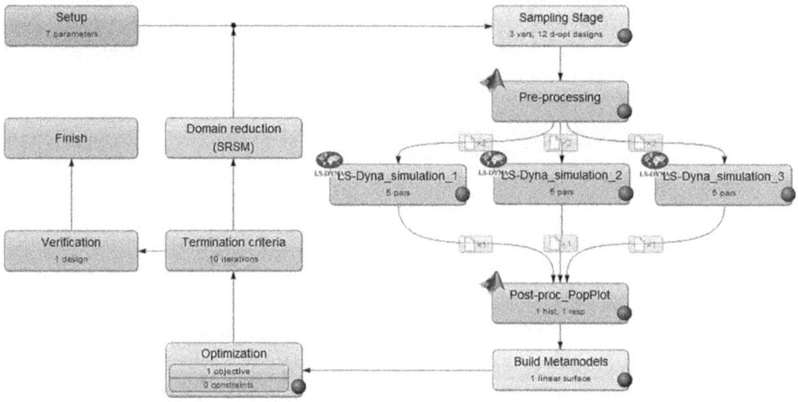

Fig. 3 LS-Opt flowchart for the evaluation of HVRB parameters.

2.2 Optimisation of parameters

In this section the results of the sensitivity study are used to setup the optimisation process to reverse engineer the parameters from directly measurable material response.

The optimisation algorithm is graphically summarised by the flowchart of the LS-Opt graphic user interface reported in Fig. 3.

The optimisation process itself consists of three stages, adding Matlab based pre- and post-processing stages to the actual simulation stage performed with LS-Dyna.

The pre-processing stage facilitates the automated parametric generation of the models, enabling the optimisation of spatial parameters (*e.g.* specimen dimensions, element size) if desired.

The post-processing stage allows the extraction of user-defined measures from the limited number of outputs available in the simulation stage, with the capability to combine the results from multiple simulations.

The work presented in this paper aims to prove the ability of the optimisation approach developed to converge onto the correct set of parameters. Therefore, the algorithm is tasked to find the baseline values presented in Table 1 using as the target only specific simulation outputs which are equivalent to directly measurable experimental results, which in this work are the Pop-plot consisting of 3 data points and the pressure profile measured at a specific gauge point along the specimen, as presented in Fig. 4.

In the following sub-sections, two sets of optimisations are presented, the first using only the Pop-plot extracted from simulations using the baseline set of parameters as optimisation target, and the second using both the Pop-plot and the pressure profile measured 0.2 mm into the energetic sample from the impact face. The accuracy and velocity of convergence is evaluated for each optimisation setup presented.

2.2.1 Single target optimisation. Given the strong correlation between the three parameters to optimise and the run to detonation distance proven in Section 2.1, the first set of simulations explores the

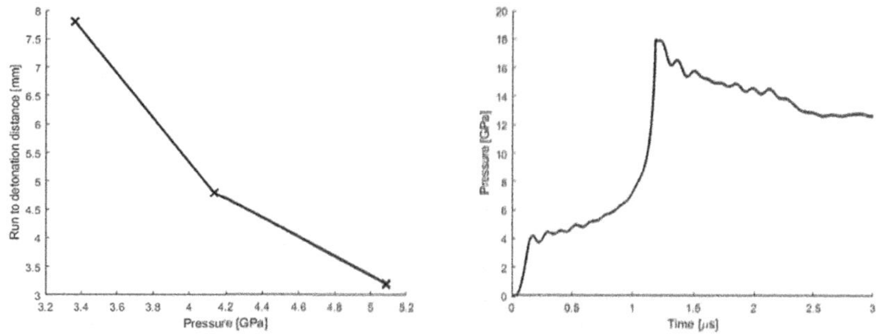

Fig. 4 SDT test outputs considered: Pop-plot (left) and pressure profile in the impacted cross-section.

Table 4 Parameter range for the optimisation study.

Parameter	Target	Parameter range	
P_r	1.0	0.75 (−25%)	1.50 (50%)
Z	1.0	0.75 (−25%)	1.50 (50%)
M	3.5	2.50 (−29%)	5.00 (43%)

capability of the algorithm to use only the Pop-plot as optimisation target.

The Pop-plot reported in Fig. 4 is extracted from the baseline model by measuring the run to detonation distance from three simulations in which the initial velocity of the impactor is imposed as 600, 700, and 800 m s^{-1}.

Imposing a parameter range consistent with the sensitivity study presented in the previous section, and reported in Table 4, the optimisation is carried out requiring 12 simulations per iteration, over 10 iterations.

The reason for the strong asymmetry of the range, beside the actual results of the sensitivity study, is that often optimisation algorithms bisect the domain to impose seeding points, therefore the algorithm might "accidentally" find the correct answer on a symmetric domain, without actually converging to it.

The results of the simulations, summarised in Fig. 5, show the ability of the optimiser to converge toward the correct solution, with six sets of parameters converging to the target Pop-plot with an objective less than 0.1. All the acceptable sets of parameters refer to simulations from the seventh iteration onwards, proving that the number of iterations imposed (*i.e.* 10) in conjunction with the domain reduction algorithm is sufficient.

However, within the six sets of parameters that nominally predict the correct HVRB parameters there is a significant variability (*i.e.* between 15% and 20%) of each of the three parameters. Comparing to the baseline set of parameters the optimiser is trying to converge to, all predictions fall within ±10% of the target values, indicating that the relations between the input parameters P_r, M, and Z and the run to detonation distance are not strongly univocal, resulting in the optimisation algorithm converging on local minima close (in parameter space) to the

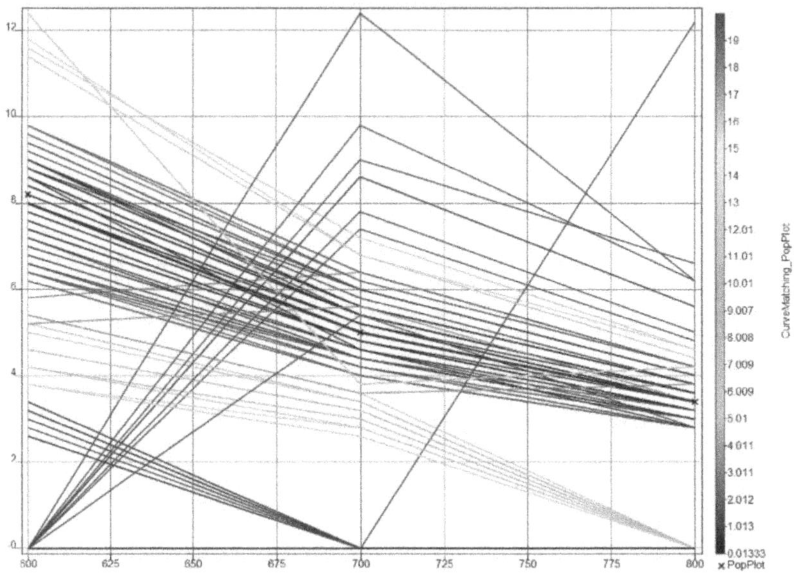

Fig. 5 Run to detonation distance *vs.* impact velocity for all the of SDT tests simulated during the single-target optimisation process (lines) against the target values (crosses).

global minimum. Hence, it can be concluded that using a single target allows to converge to values close (*i.e.* ≤10%) to the correct answer, but the correlation is not strong enough for the fine tuning of the parameters.

2.2.2 Multi-target optimisation. To improve the accuracy of the optimisation algorithm, a second convergence target (*i.e.* pressure profile on the impacted face, graphically presented in Fig. 4) is introduced. The same range of parameters as of the single-target optimisation, summarised in Table 4, is applied, leading to a significant improvement of the level of convergence, with the most accurate prediction of the run to detonation distance converging to the exact target values (*i.e.* objective $\leq 10^{-4}$), due to all three HVRB parameters converging to the target (baseline) values with an error of less than 2%.

Compared with the single target optimisation, the added target does not affect the speed of convergence, as the results again become acceptable from the seventh iteration. However, the correlation matrix reported in Fig. 6 shows graphically the reason for the shortcoming of the single-target optimisation reported in the previous section, with both P_r and Z showing weaker correlations to the run to detonation distance than to the pressure profile. The opposite can be said about M, demonstrating the need for both objectives to be imposed concurrently.

The ability to univocally converge towards the correct set of parameters demonstrates the validity of the overall approach developed, with the combination of an initial sensitivity study and subsequent multi-target optimisation process, proving the potential of the optimisation framework implemented to reverse-engineering HVRB parameter sets using target curves directly measured from real-world experimental SDT tests.

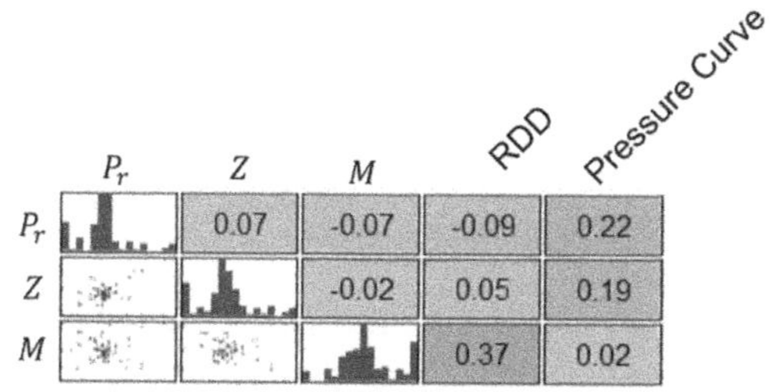

Fig. 6 Correlation matrix of each HVRB parameter against the other parameters and the simulation outputs, *i.e.* run to detonation distance (RDD) and the pressure profile.

Table 5 Reduced and increased parameter ranges to study the effect of the domain size upon the optimisation outcome.

Parameter	Target	Small parameter range		Large parameter range	
P_r	1.0	0.90 (−10%)	1.15 (15%)	0.25 (−75%)	2.0 (100%)
Z	1.0	0.90 (−10%)	1.15 (15%)	0.25 (−75%)	2.0 (100%)
M	3.5	3.15 (−10%)	4.00 (14%)	1.00 (−71%)	7.0 (100%)

Table 6 Comparison of optimisation accuracy for three different parameter domain sizes.

Range set	P_r	Z	M
Small	0.96 (−3.9%)	1.01 (1.5%)	3.44 (−1.8%)
Regular	1.02 (1.8%)	1.02 (1.7%)	3.56 (1.8%)
Large	0.85 (−14.9%)	1.37 (37.3%)	1.00 (−2.3%)

2.3 Verification of the parameter range

Finally, the multi-target optimiser is used to explore the effect of the parameter range on the optimisation process, in order to verify the validity of the sensitivity study presented in Section 2.1.

Two additional sets of parameter ranges, one smaller and one larger than the one reported in Table 4, are defined (reported in Table 5) and used as setup for the optimisation process as presented in Section 2.2.

The outcome of the simulations on the small parameter range set is that reducing the range accelerates the convergency process, with a level of agreement similar to the one presented in Section 2.2 reached by the fourth iteration rather than the seventh. However, no improvement is observed in the accuracy of the prediction of the HVRB parameters, with the same ≈2% difference with respect to the target values being observed.

Imposing a larger parameter range, beyond the bounds defined in the sensitivity study, instead, significantly affects the quality of the optimisation, leading to very large error in the prediction of the HVRB parameters, with errors exceeding 30% even for the nominally optimal solution, as summarised in Table 6.

The inability of the algorithm to correctly predict the target parameters is due to the impossibility to differentiating between simulations in the overdriven and underdriven regions, which does not allow the determination of a gradient for the domain reduction outside the parameters' bounds determined in Section 2.1. The outcome of the optimisation process on an excessively large domain, therefore, further proves the necessity of a comprehensive sensitivity study.

3 Conclusions

This paper demonstrated the development and implementation of a numerical framework to evaluate HVRB parameters using the experimentally measurable response of explosives, using numerical simulations (with Lagrangian formulation) of the shock-to-detonation transition (SDT) test. A systematic sensitivity study is performed to evaluate the relevant parameters for the test, the reasonable range of each parameter, and the relevant outputs to use as convergence targets for the subsequent optimisation study. The optimisation algorithm itself consists of three stages, a pre-processing stage for the parametric generation of the FE model, a simulation stage relying on the commercial software LS-Dyna, and a post-processing stage for the extraction of outputs.

The studies presented show that, despite the run to detonation distance being the output most sensitive to parameter variation, its use alone as an objective is insufficient to achieve accurate and unique parametrisation. The simultaneous use of run to detonation distance and pressure profile as objectives provides an acceptable level of constraint to accurately resolve the HVRB parameters.

Finally, the effect of the parametrisation range was explored showing that the reduction of the domain leads to quicker convergence towards the correct values, without necessarily increasing the accuracy. The use of excessively large domains instead (*i.e.* beyond the bounds for each parameter defined in the sensitivity study), can significantly affect the accuracy of the prediction, further demonstrating the importance of conducting sensitivity studies ahead of the optimisation processes.

Acknowledgements

The authors gratefully acknowledge the support of the United Kingdom Defence Science and Technology Laboratory (DSTL), Grant No. DSTLX1000144542.

References

1 G. I. Kerley, *CTH Equation of State Package: Porosity and Reactive Burn Models*, Sandia National Laboratories report SAND92-0553, 1992.
2 S. Falco, D. Chapman, A. Tyas, A. Limbu, M. Whittaker, D. Pope, N. Petrinic, and D. Eakins, Implementation of HVRB in LS-Dyna and application to detonation in hemispherical charges in ideal explosives, in *Future Developments in Explosives and Energetics: 1st International Explosives Conference*, Royal Society of Chemistry, 2023, p. 220.
3 J. M. McGlaun, S. L. Thompson and M. G. Elrick, CTH: A three-dimensional shock wave physics code, *Int. J. Impact Eng.*, 1990, 351.

4 J. O. Halloquist, LS-Dyna theory manual, Livermore Software Technology Corporation, 2006.
5 J. Starkenberg and T. M. Dorsey, *An assessment of the Performance of the History Variable Reactive Burn Explosive Initiation Model in the CTH code*, Army Research Lab Aberdeen Proving Ground MD Weapons Technology Directorate, ARL-TR-1667, 1998.
6 Y. Wei, C. Miller, D. Olsen and M. Zhou, Prediction of propabilitic shock initiation threshold of energetic materials through evolution of thermal-mechanical dissipation and reactive heating, *J. Appl. Mech.*, 2021, **88**, 091005.

Analysis of the burning of spheres and cubes†

John Philip Curtis,*[a,b] Adam Schofield[a] and William Belfield[a]

Modellers of explosive burn typically consider the inward burning of a spherical grain, uniformly ignited at its surface. Recent work addressed the inward burning of other regular shapes, assuming a constant burn speed and gave support to using a spherical grain shape. However, internal ignition of the explosive grain at a hotspot causes outward burning, the subject of this paper. First addressed is the ignition of a solid sphere ignited at its centre. This produces a spherical burn front, moving at constant speed until it reaches the surface when the sphere is fully burnt. The burn fraction as a function of distance travelled is compared with the inward burn for the same size of sphere. Next, the burn of a solid cube ignited at its centre is considered. The burn fraction at an arbitrary time is determined by performing the appropriate volume integrals. This fraction as a function of the distance travelled by the front is compared with other shapes of equal volume. It is shown that Monte Carlo integration can accurately predict the burn fraction. This method shows promise for the consideration of the burn of arbitrary shapes. Potential developments of this work are discussed.

1 Introduction

Modelling the burn process is important as a tool for the assessment of explosive safety and the prediction of the consequences of accidents, caused by thermal and mechanical hazards. There often remains uncertainty in determining whether the ignition of an explosive will cause a violent reaction. For a given burn velocity, the rate at which the solid explosive produces gas product increases with the amount of surface area that has been ignited. As a result, in a fixed volume, the rate of pressure increase also depends on that surface area. Since, in general, the burn velocity increases with pressure, the build-up to a violent response can accelerate dramatically. Depending on the confinement, the response can be a violent burn, possibly leaving some of the explosive unconsumed, or a detonation, which typically consumes all the available explosive. Previous work with the High Explosive Response to MEchanical Stimulus (HERMES) model by Reaugh *et al.*[1] has explored the considerable dependence of the explosive response on the Specific Surface Area (SSA).

The process of a burn gathering in severity to grow to detonation is known as Deflagration-to-Detonation Transition (DDT). The importance of DDT has engendered a considerable body of modelling work. It is beyond the scope of this paper to review all the literature in this field, but it is appropriate to draw the reader's attention to certain work that the

[a] *AWE Aldermaston, Reading, Berkshire RG7 4PR, UK. E-mail: John.P.Curtis@awe.co.uk*
[b] *University College London, Gower Street, London WC1E 6BT, UK. E-mail: John.Curtis@ucl.ac.uk*

present work is intended to complement in future. Saenz and Stewart[2] have commented that 'macroscale DDT in explosive powders has well-defined characteristics and is amenable to description by continuum models that in some sense average over the microstructure'. They recognised the challenge of keeping a computational model simple enough to be practical regarding resource usage, while nonetheless capturing the essential physics. They summarise modelling effort by other researchers, including the fundamental work by Baer and Nunziato[3] in which a truly multi-phase continuum description was offered. Curtis[4] has developed analysis to generalise this work to include strength effects. The philosophy of the HERMES model[1] has also been to rise to the challenge of retaining simplicity and usability while capturing the physics.

In their modelling of the DDT experiments of Sandusky *et al.*,[5] which used a confined cylinder of explosive powder, Miller *et al.*[6] considered spherical particles. They applied log-normal distributions of the radii of LX04 powder explosives of a few mm in diameter. By fitting numerical integrations over the distributions, they were able to represent the burn-up behaviour with a single equivalent particle radius, with matched SSA. The observed DDT behaviour could then be modelled well by making reasonable estimates of shock-to-detonation parameters. This was encouraging, but the semi-empirical nature of the model was acknowledged. The very special nature of the spherical geometry was recognised – the sphere has the least surface area for given volume and this special property may have significance in addressing burning. More recently, Reaugh *et al.*[7] applied the same methods as used previously to investigate DDT in the different explosive PETN, obtaining good agreement with the experimental work of Luebcke *et al.*[8]

DDT can also occur in Polymer-Bonded Explosives (PBXs). These explosives comprise explosive crystals embedded in a polymer matrix, sometimes itself reactive. At the mesoscopic scale typical of PBXs, the crystal shapes are usually not spherical. This naturally prompted the question 'How would the burning behaviour change if the particle shapes were of other than spherical shape?' It is worthy of note that in PBXs, correlation between the SSA of the crystals and the sensitivity of the explosive has been postulated. SSA changes with length-to-diameter (L/D) ratio for shapes such as a cylinder, for example, and so one may infer that shape must affect burn rate.

In a preliminary investigation of whether grain shape is indeed important, previous work[9] addressed the inward burning of some common regular shapes, assuming a constant speed of the burn propagation. It was considered that this would give useful insight into the effects of geometry, though more realistic treatment of burning should then follow. The approach was thus a simplifying attempt to isolate shape effects. Results for the selected shapes, namely cubes, regular tetrahedra, and cuboids, were compared and were not hugely different in nature provided the values of L/D stayed near unity. The work thus gave some support to the choice of the sphere as the grain shape. However, as expected, the results did differ somewhat. Notable was the increased rapidity of burning of cuboids with increasing length-to-diameter ratio. It was acknowledged that in some real situations the explosive grain is ignited at some internal hotspot and the burn progresses outwards from this localised heat source. The recommendation was made that the research of such scenarios be undertaken.

Accordingly, after a summary of the inward burning results,[9] this paper addresses such outward burning for some very fundamental geometries. It is found that even these can offer non-trivial analytical challenges. First, perhaps the most fundamental scenario of the burning of a solid sphere of explosive ignited at its centre is considered. It is assumed that the outward burn progresses as a spherical front moving at constant speed until the entire solid sphere is consumed. The burn fraction as a function of distance travelled is compared with the inward burn for the same size of sphere. Next, for what seems another fundamental scenario of great interest, the burn of a solid cube ignited at its centre is considered. The burn front is again a sphere of increasing radius and the burn fraction at an arbitrary time is determined by performing the appropriate volume integrals analytically. This is straightforward until the front radius increases to the point where the cube no longer remains intact, with only the eight corners left, when the evaluation of the unburnt volume left is non-trivial. Eventually, the burn front consumes the entire cube. The burn fraction as a function of distance travelled by the front is compared with other shapes of the same volume. It is shown that Monte Carlo integration with a sufficiently high number of trials accurately predicts the burn fraction. This method shows promise for the consideration of the burning of arbitrary shapes, where other analytical or numerical methods may be cumbersome. The paper concludes with a discussion of how the implementation of more realistic burn models may be accomplished and then suggests some extensions and generalisations of this work.

2 Summary of inward burning results

Based on the assumption of a constant burn speed in previous work[9] comparisons were made between the burn rates and evolution of specific surface area (SSA) associated with spheres, cubes, tetrahedra and cuboids of square cross-section having $L/D > 1$, all having the same volume. All the dimensions of each shape were expressed in terms of the sphere radius. The sphere, having the least initial specific surface area for the prescribed volume, burns the slowest while the $L/D = 10$ cuboid of square cross-section is the fastest burning shape of those investigated. Although the study was confined to a limited set of shapes, it was reasonable to conclude that explosives containing a significant proportion of crystals of high L/D shapes, of whatever cross-section, or having greater SSA for other reasons such as irregular shapes, will burn more rapidly than those having a predominance of regularly shaped crystals, for the same chemical species. There are likely to be correlations with increased sensitivity and increased violence associated with the burn since the high-pressure products are generated faster.

For convenience, the burn fractions as a function of time are restated here for the inward burning of the sphere and of the cube. The reader is referred to the previous work[9] for the other shapes. For the sphere, the burn fraction f is given as a function of time by

$$f = 1 - \frac{(r_0 - v_B t)^3}{{r_0}^3} = 1 - \frac{(r_0 - x(t))^3}{{r_0}^3} \tag{1}$$

Here r_0 is the initial radius of the sphere, t is time, v_B is the constant burn speed, inward in this case, and $x(t)$ is the run distance of the burn, which it is convenient to introduce for later comparisons of the burn models. Hence burning is fully complete ($f = 1$) at time t_C and run distance $x(t_C)$ given by:

$$t_C = \frac{r_0}{v_B}, \quad x(t_C) = r_0. \tag{2}$$

The expression for the inward burning of the cube is similar. It is

$$f = 1 - \frac{(a_0 - 2v_B t)^3}{{a_0}^3} = 1 - \frac{(a_0 - 2x(t))^3}{{a_0}^3}. \tag{3}$$

Here a_0 is the initial side length, t and v_B are as defined for the sphere, and $x(t)$ is the run distance of the burn, with the difference that this is measured normal to the faces of the cube and thus not measured towards the centre of the sphere. The cube stays a cube as the burn progresses in this simple treatment. Some rounding as the burning proceeds would be expected at the corners, and this effect should be investigated in future work. However, here the burn is fully complete at time t_C and run distance $x(t_C)$ given by:

$$t_C = \frac{a_0}{2v_B}, \quad x(t_C) = \frac{1}{2}a_0. \tag{4}$$

Comparisons with these results for inward burning are made later in this paper.

3 Outward burning of a sphere

The outward burning of a sphere of radius r_0 is now addressed. Again, the constant burn speed is v_B. The run distance $x(t)$ is now of course measured from the centre of the sphere. The volume $V(t)$ burnt at time t is given by

$$V(t) = \frac{4}{3}\pi {v_B}^3 t^3 = \frac{4}{3}\pi x^3(t). \tag{5}$$

The burn fraction is thus very simply given by

$$f = \frac{{v_B}^3 t^3}{{r_0}^3} = \frac{x^3(t)}{{r_0}^3}. \tag{6}$$

Burning is completed at the time and run distance given by eqn (2), as for the inward burn. Eqn (6) may be compared with eqn (1) and results are shown in Fig. 2. It is immediately apparent that much more material is rapidly burned at early times for the inward burn, while more is burned at later times for the outward burn. This reflects the much greater initial specific surface area for the inward burn. In this simple analysis there is no change in the time taken to reach complete burn. However, in view of the role of product pressure reinforcing the speed of burn, this geometric change is likely to have significant consequences for burn time.

4 Outward burning of a cube

The outward burning of a cube of edge length a_0 ignited at its centre is now considered. The perpendicular distance from the centre to any face is $a = \frac{1}{2}a_0$.

The distance from the centre to the midpoint of an edge is $\sqrt{2}a$ and to a corner is $\sqrt{3}a$.

It is assumed that the burn progresses as an expanding sphere with the burn front again moving at speed v_B, with run distance $x(t) = v_B t$. The burn progresses in three stages.

For $t < \frac{a}{v_B}$ the spherical burnt region lies entirely inside the cube, as shown in Fig. 1(a). For $\frac{a}{v_B} \le t < \frac{\sqrt{2}a}{v_B}$ the burnt region has reached the side faces of the cube but not the edges, as shown in Fig. 1(b). The unburnt material in the cube has been holed but remains connected. For $\frac{\sqrt{2}a}{v_B} \le t < \frac{\sqrt{3}a}{v_B}$ the burnt region has reached the edges and the unburnt material in the cube is no longer intact – just regions of unburnt material adjacent to the corners survive, as seen in Fig. 1(c). Finally, the cube is fully burnt at time $t = \frac{\sqrt{3}a}{v_B}$.

The complexity of evaluating the burn fraction increases with each successive stage of the burn. For the first stage where the burnt sphere is entirely within the cube it is simply given by the volume of the full sphere divided by that of the cube, namely,

$$f = \frac{\frac{4}{3}\pi {v_B}^3 t^3}{8a^3} = \frac{\pi {v_B}^3 t^3}{6a^3} = \frac{\pi x^3(t)}{6a^3} \quad \text{for} \quad r < a, t < \frac{a}{v_B}, \tag{7}$$

where $x(t)$ is again the run distance. For the second stage where the burn has reached the faces, but not the edges of the cube, the amount burnt

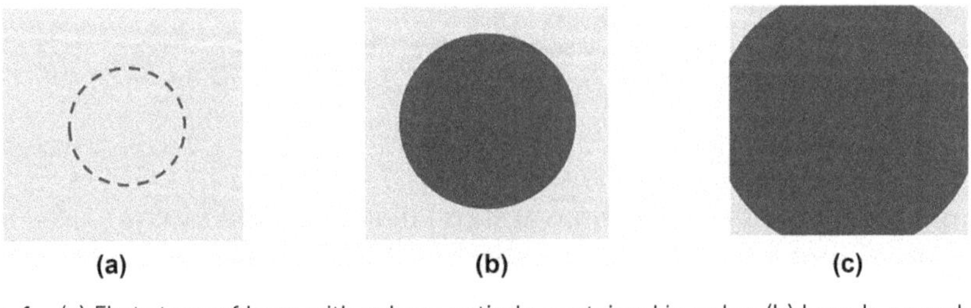

Fig. 1 (a) First stage of burn with sphere entirely contained in cube; (b) burn has reached the side faces; (c) burn has reached the edges.

may be calculated as the volume of the full sphere, minus the six spherical caps with bases defined by the intersections of the sphere with each face. The volume of each spherical cap is found by calculating the volume of the associated spherical sector minus the cone defined by the base and the centre of the cube. Hence the volume of the burnt region is

$$V(t) = \frac{2\pi}{3}\left(9r^2a - 3a^3 - 4r^3\right) \quad \text{for} \quad a \leq r < \sqrt{2}a. \tag{8}$$

The burn fraction for stage 2 is thus:

$$f = \frac{\pi}{12a^3}\left(9x^2(t)a - 3a^3 - 4x^3(t)\right) \quad \text{for} \quad \frac{a}{v_{\mathrm{B}}} \leq t < \frac{\sqrt{2}a}{v_{\mathrm{B}}}, \tag{9}$$

where, as before, $x(t) = v_{\mathrm{B}}t$.

The third burn stage posed a considerably greater challenge. The calculation of the burnt volume includes the volume of a cuboid of height $2\sqrt{r^2 - 2a^2}$ and two sides of the cube of length $2a$, plus two integrals with respect to the z-Cartesian co-ordinate, with the origin of the co-ordinate system at the centre of the cube and each axis parallel to four edges of the cube. In the regions $\sqrt{r^2 - 2a^2} \leq z \leq a$ and its symmetric counterpart $-a \leq z \leq -\sqrt{r^2 - 2a^2}$ the intersection of the sphere and the z-plane are four arcs of a circle of radius $\sqrt{r^2 - z^2}$. These subtend four circular sectors centred on the z-axis in the z-plane. The remainder of the burnt cross-section in the z-plane comprises four triangles of base length $2\sqrt{r^2 - z^2 - a^2}$ and height a.

Thus, the burnt volume V can be expressed after some manipulation as

$$\begin{aligned} V &= 8a^2\sqrt{r^2 - 2a^2} + 8\int_{\sqrt{r^2 - 2a^2}}^{a} \\ &\quad \times \left\{a\sqrt{r^2 - z^2 - a^2} + (r^2 - z^2)\left(\frac{\pi}{4} - \cos^{-1}\left(\frac{a}{\sqrt{r^2 - z^2}}\right)\right)\right\}\mathrm{d}z. \end{aligned} \tag{10}$$

Further manipulations and integrations yield:

$$\begin{aligned} V &= 8a^2\sqrt{r^2 - 2a^2} + \frac{16a}{3}(3r^2 - 2a^2)\sin^{-1}\left(\frac{a}{\sqrt{r^2 - a^2}}\right) + \frac{4\pi a}{3}(-3r^2 + 2a^2) \\ &\quad - \frac{8a}{3}\int_{\sqrt{r^2 - 2a^2}}^{a} \frac{z^2(3r^2 - z^2)}{(r^2 - z^2)\sqrt{r^2 - z^2 - a^2}}\mathrm{d}z. \end{aligned} \tag{11}$$

With some effort the final term may be shown to be

$$-\frac{4a^3}{3}\begin{bmatrix} (3k^2 + 1)\left\{\tan^{-1}\left(\sqrt{k^2 - 2}\right) - \tan^{-1}\left(\frac{1}{\sqrt{k^2 - 2}}\right)\right\} \\ +4k^3\left\{\tan^{-1}\left(\frac{1}{k\sqrt{k^2 - 2}}\right) - \tan^{-1}\left(\frac{\sqrt{k^2 - 2}}{k}\right)\right\} \end{bmatrix}, \tag{12}$$

where $k = \frac{r}{a}$. The burn fraction for stage 3 is

$$f = \frac{V}{8a^3}, \quad \text{for} \quad \frac{\sqrt{2}a}{v_B} \leq t < \frac{\sqrt{3}a}{v_B}, \tag{13}$$

where this time V is given by eqn (11) using (12) to replace minus the integral term and

$$r = x(t) = v_B t. \tag{14}$$

5 Comparison of the inward and outward burning of spheres and cubes

The dependence of burn fraction on run distance for the inward and outward burning of spheres and cubes may now be directly compared by using eqn (1), (3), (6) and (13) (using (11), (12) and (14) in support). It is convenient to work with run distance rather than time as time plots will be dependent on the particular burn speed v_B and the analysis in terms of run distance applies for all burn speeds. For the same volume of material in both sphere and cube the run distance is conveniently expressed in terms of a multiple $\bar{x}$ of half the length of the edge of the cube, *i.e.*, a. The plot of burn fraction f against $\bar{x}$ is then independent of burn speed and dimensions of the sphere and cube, so applies to all pairs of spheres and cubes having the same volume burning at any burn speed.

The sphere having the same volume as the cube has radius $1.24070a$ (to 5 d.p.). This is the final run distance for completion of both inward and outward burning of the sphere. For inward burning of the cube, the burn is always normal to the face and so the final run distance is simply a. For outward burning of the cube the final run distance is $\sqrt{3}a = 1.73205a$.

Fig. 2 shows the dependence of the burn fraction as a function of run distance for the inward and outward burning of both the sphere and the

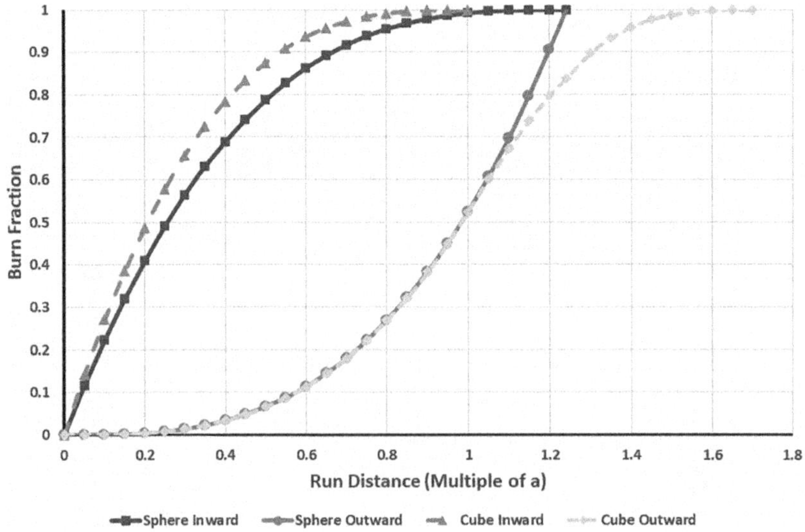

Fig. 2 Burn fraction as a function of run distance expressed in multiples of half of the edge length, *a*.

cube. The figure shows that the plots of burn fraction for inward burn are concave downwards for both the sphere and the cube. Higher burn fractions are achieved for lower run distances. As reported in ref. 9, the cube burns faster than the sphere. For the outward burn this situation is reversed. The sphere is fully burnt well before the cube. Until $\bar{x}=1$, in these models there is no difference in the burn fraction curves for the sphere and cube. At that point the burn fraction rate of increase with $\bar{x}$ continues to increase for the sphere – more and more material is burned until the outer radius of the sphere, namely $1.24070a$, is reached. The rate of increase of the burn starts to decrease at $\bar{x}=1$ for the cube, *i.e.*, when the sphere meets the faces of the cube. The burning of the remaining portion of the cube proceeds like the inward burns as the available surface area diminishes.

The spherical symmetry ensures that completion of the burn occurs at the same distance in the inward and outward burns of the sphere, owing to the assumption of constant ν_B made in this analysis. However, the more rapid inward burn would be causing faster temperature and pressure increases in the products and these would probably speed up the burn rate even more. The improvement of the physics, chemistry and geometric realism of this model will be discussed in a later section. Nonetheless, this simple model, albeit based on perhaps the simplest possible shapes to consider, with the simplest possible ignition scenario of centre ignition, has already suggested that the inward burning of crystals or grains is likely to be more rapid than outward burning from some interior hotspot.

6 Monte-Carlo evaluation of the burn fraction

The derivation of eqn (11) and (12) was algebraically heavy and it took some time to achieve confidence in the above results. These difficulties

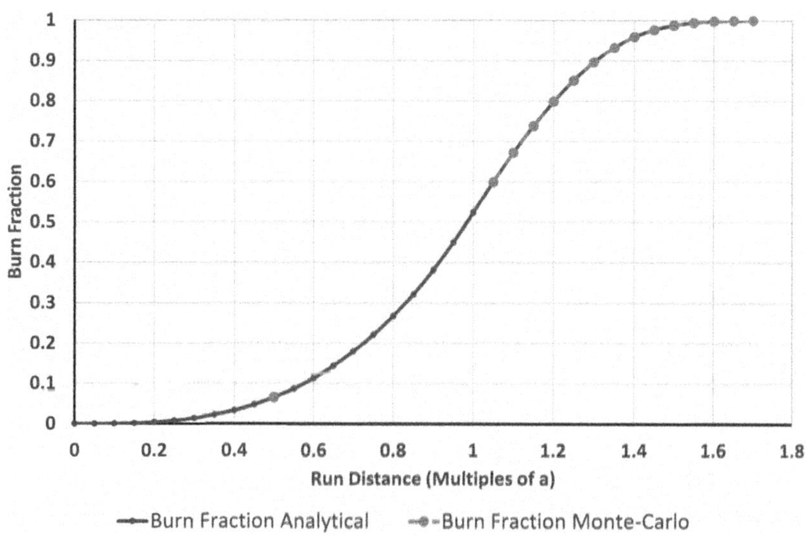

Fig. 3 Burn fraction as a function of run distance for the outward burning of a cube, expressed in multiples of half of the edge length, *a*. The analytical result is compared directly with the Monte-Carlo method.

prompted the parallel approach, in which Monte Carlo integration was applied to the same integration for the outward burning of the cube, covering all the three stages identified above. This method is more concisely coded than the analytical method described above, the reduction in the effort on analysis coming at the cost of performing many Monte Carlo trials.

The analytical and Monte-Carlo method results are directly compared for the outward burning cube in Fig. 3. It may be seen that the agreement for 10 000 trials was excellent.

7 Discussion

As yet, only symmetric inward and outward burns have been addressed assuming constant speed of burn. For the outward burns point ignition at the centres of the spheres and cubes has been assumed. The analysis for even these most basic of shapes has already revealed some very interesting differences in the dependence of burn fraction upon grain shape and the location of ignition. Increasing and decreasing rates of burn are associated with outward and inward burning respectively. As Fig. 2 very clearly shows, inward burn is much more rapid and in practice would raise pressure and temperature of larger volumes of explosive more quickly, accelerating the burn even faster than the constant burn speed model can address. It is expected that the results may change significantly as the location of the ignition point changes, even for constant burn speeds. In general, one is likely to have to address arbitrary ignition locations, which may occur at inner cavities inside the explosive. Such cavities may themselves be of irregular shape. The explosive grain will never be of perfect geometrical shape, unless three-dimensional printing and additive manufacturing methods advance to such a stage that this becomes feasible. Then solutions for ideal geometrical shapes may acquire new importance. Based on models of the effects of pinch and shear previously investigated[10,11] there may then be exciting possibilities for the precise control of the sensitivity of explosive configurations to likely insults.

Remaining for now with the assumption of constant burn speed, the calculation of the burnt region for the diversity of irregular-shaped grains seen in real explosives will pose very challenging volume integrations, likely to be infeasible within reasonable timescales using analytical methods akin to those presented in this paper. Even the tetrahedron and cuboid, for which inward burn was investigated[9] pose some challenging analysis. Numerical integration methods could possibly be applied, but the diversity of shapes and initiation possibilities renders this likely to be infeasible for arbitrary configurations. The authors believe that Monte-Carlo integration applied to a representative sample of explosive where the shapes of the grains and cavities are found by tomography or other methods, may be the key to obtaining at least a good approximation to the burn fraction history of a multi-grain explosive with arbitrary porous regions and grain shapes. Progress towards exploring this idea further is described in a sister paper at this conference.[12]

The main aim of this paper was to undertake an investigation of the effects of geometry upon burn. This investigation is believed to have demonstrated the significance of geometry. However, the findings with the constant burn speed have if anything emphasised the need to improve on the burn model physics. Dependence on the high pressures and temperatures generated as the burn proceeds must surely feature in a physically realistic treatment. How may this be accomplished?

Previous works by Catalano *et al.*,[13] Breshears,[14] Zerkle,[15] and Curtis[16,17] have modelled the reaction of a sphere of explosive in the One-Dimensional Time to Explosion (ODTX) experiment using Arrhenius models for the chemical reactions. Curtis[17,18] adapted his work to address the heating of a slab, applying it to the heating of an explosive by a gas being compressed. Said *et al.*[19,20] have investigated the interaction of hot gas with explosive, allowing the hot gas to feedback heat to the explosive. There is potential to include more of the required physics by exploiting this research. It seems well worthwhile to explore whether these methods can be combined with the Monte-Carlo approach to calculating the overall burn fraction.

Future work could address the further investigation of both regular and arbitrary solid shapes with arbitrary ignition sites. Also of interest is the study of the effects of hollow shapes ignited at the surface of the hollow. Previous work on the reaction of spheres and slabs and of gas product heating, or generalisations thereof, should enable more realistic models for burn front speeds. Importantly, further investigation of the Monte-Carlo method and its combination with the above areas should be undertaken.

References

1 J. E. Reaugh, B. W. White, J. P. Curtis and K. Springer, *Propellants, Explos., Pyrotech.*, 2018, **43**, 703.

2 J. A. Saenz and D. S. Stewart, *J. Appl. Phys.*, 2008, **104**, 043519.

3 M. R. Baer and J. W. Nunziato, *Int. J. Multiphase Flow*, 1986, **12**, 861.

4 J. P. Curtis, *A One-Dimensional Formulation of a Two-Phase Continuum Model with an Ongoing Reaction*, Unpublished MoD Report.

5 H. W. Sandusky, R. H. Granholm, D. G. Bohl, K. S. Vandersall, D. E. Hare and F. Garcia, *Proceedings of the 13th International Detonation Symposium*, Norfolk, VA, USA, 2006.

6 C. M. Miller, B. W. White and J. E. Reaugh, *Proceedings of the 22nd Biennial Conference of the APS Topical Group on Shock Compression of Condensed Matter*, Anaheim, CA, USA, 2022, vol. 67.

7 J. E. Reaugh, J. P. Curtis, and M.-A. Maheswaran, *Proceedings of the 15th International Detonation Symposium*, San Francisco, California, USA, 2014.

8 P. E. Luebcke, P. M. Dickson and J. E. Field, *J. Appl. Phys.*, 1996, **79**, 3499.

9 J. P. Curtis, J. E. Reaugh, and C. A. Handley, *Europyro 2019, 44th International Pyrotechnics Society Seminar*, Tours, France, 2019.

10 J. P. Curtis, *Proceedings of the Thirty-Eighth International Pyrotechnics Seminar*, Denver, Colorado, 2012, p. 207.

11 J. P. Curtis, *Proceedings of the 39th International Pyrotechnics Seminar*, Valencia, Spain, 2013.

12 A. Schofield, W. Belfield, and J. P. Curtis, *Proceedings of the 2nd International Explosives Conference*, Edinburgh, 2024.
13 A. E. Catalano, R. McGuire, E. Lee, E. Wrenn, D. Ornellas and J. Walton, *Sixth Symposium (International) on Detonation*, Coronado, California, USA, 1976.
14 D. Breshears, *One-Dimensional Time-to-Explode (ODTX) in HMX Spheres, Los Alamos Internal Report*, LA-UR-98-1237, 1998.
15 D. H. Zerkle, *Multiphase Treatment of ODTX in HMX Spheres, Los Alamos Internal Memorandum*, LA-UR-98-1238, 1997.
16 J. P. Curtis, *A New Analytical Model of the One-Dimensional Time to Explosion Experiment*, Unpublished MoD Report.
17 J. P. Curtis, *Refinement of the One-Dimensional Time to Explosion and Slab Reaction-Diffusion Model*, Unpublished MoD Report.
18 J. P. Curtis, *Proceedings of the 1st International Explosives Conference*, ed. J. Akhavan, Royal Society of Chemistry, London, 2022, p. 232.
19 S. Said, PhD thesis, University College London, 2022.
20 S. Said, F. T. Smith and J. P. Curtis, Modelling, computation and analysis on combustion of explosives, *Eur. J. Appl. Math.*, 2021, **33**(1), 27–57.

Analysis technique for reconstructing 2D cylinder wall expansion†

Thomas Scott

A cylinder test consists of detonating an explosive encased by a metal cylinder and then recording the expansion of the cylinder using Heterodyne Velocimetry (HetV) probes. The HetV probes record the velocity–time profile of the point on the cylinder wall that the probe views, but no spatial information. In the AWE design, ten HetV probes are mounted perpendicularly to the cylinder wall. As only a component of the wall motion is recorded by these probes, integrating the velocity–time profile recorded by these probes is not equal to the cylinder wall motion. Therefore, three probes are fielded at 20, 30 and 40 degrees respectively to the cylinder wall to record multiple components of the total wall velocity. Due to the geometry of the experiment, the angled recordings are not linked in space and/or time and a reconstruction method is required to convert the probe recordings into the motion of a single particle on the cylinder wall. This paper proposes a new approximate technique that uses the data from a probe fielded perpendicularly to the cylinder wall and another fielded at a known angle. This technique is demonstrated on cylinder test data from an HMX-based explosive.

1 Introduction

The cylinder test is a common explosive performance test that can be used to determine the Jones–Wilkins–Lee equation of state of an explosive.[1] However, due to the experimental geometry, it is not possible to record the data required for this analysis from a singular velocimetry probe, so multiple measurements must be combined to reconstruct the data required.[2–4] This paper discusses a novel approximate method to perform this reconstruction.

2 Experimental geometry

A cylinder test consists of explosive cylinder encased by a copper tube that is initiated by a detonator at one end face that causes a detonation wave to travel from the initiation end down the central axis of the explosive cylinder to the opposite end. Once the detonation wave has travelled through sufficient explosive to stabilise (~6–8 charge diameters long), the detonation wave will be in steady-state, and uniformly travel along the central axis of the explosive cylinder with a detonation velocity that is material- and diameter-dependent. As the detonation wave travels along the explosive cylinder, it will generate a shock wave in the confining copper cylinder. This shock wave will travel through the copper cylinder and once it reaches the cylinders outer surface, the shock wave will cause the cylinder wall to move outwards. The velocity that an individual particle on the cylinder moves with will be called the wall velocity, and if the detonation wave is in steady-state, the wall

AWE Nuclear Security Technologies, UK. E-mail: Thomas.Scott@awe.co.uk

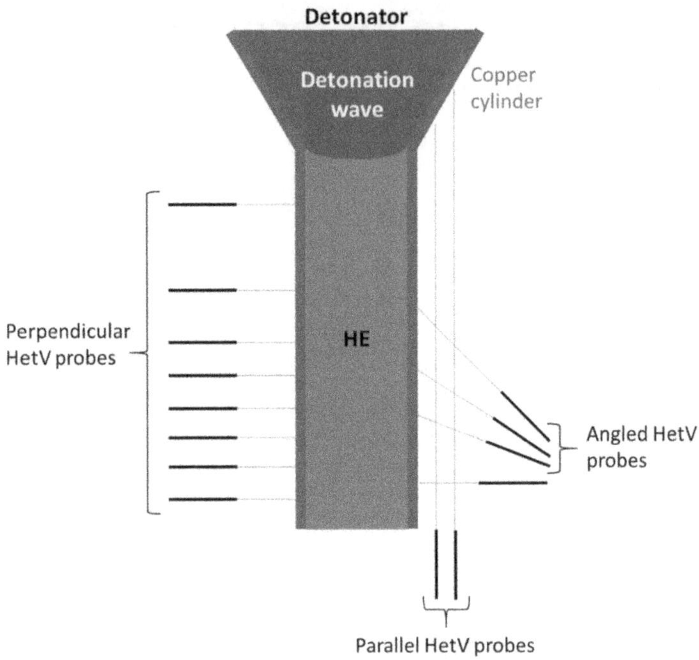

Fig. 1 Illustration of a typical cylinder test setup.

velocity–time profile experienced by each particle on the cylinder wall will be the same.

The velocity of the expanding copper cylinder wall in steady-state is recorded using an array of Heterodyne Velocimetry (HetV) probes.[5] The HetV probes record the wall velocity towards the probe of the point on the cylinder wall that the probe views, but no spatial information. Typically, twelve HetV probes are fielded perpendicularly to the cylinder wall, three at an angle to the perpendicular (nominally 20, 30 and 40 degrees) and two parallel to the cylinder wall, which is illustrated in Fig. 1.

3 Theory

Due to the experimental geometry, each HetV probe records a different particle compared to each other which changes with time. To explain why this occurs, a simplified setup will be considered that consists of three HetV probes that view the copper wall in steady-state conditions: one perpendicular to the cylinder wall and two at equal and opposite angles called A and B. The three probes are aligned so that they initially view the same particle, called the alignment particle, on the cylinder wall. This setup is shown in the top image in Fig. 2. The probes record continuous data that start at time zero, when the alignment particle starts moving, and end when the cylinder wall in the region the probe views breaks apart.

After an arbitrary unit of time from time zero, the detonation wave moves with the detonation velocity from the alignment particle along the explosive cylinder axis. Each particle in the cylinder wall it passes is given

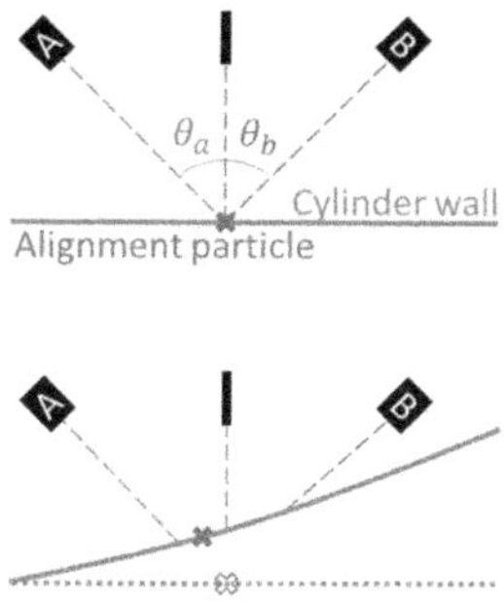

Fig. 2 [top] Simplified cylinder wall and HetV setup at time zero with the angle of angled probe A and B labelled. [bottom] Setup after one arbitrary time step with previous cylinder wall and alignment particle position shown.

energy by the expanding product gases and moves out from the cylinder axis at an angle with the wall velocity. The motion of the alignment particle was not in the direction of any of the probes, so none of the probes are recording the particle alignment. Each probe now records a different particle compared to each other. As the wall expansion is steady-state, each particle is following the same wall velocity–time relationship but each particle has expanded a different amount of distance/time. In this example, both the perpendicular probe and angled probe B and view a particle that has moved further than the alignment particle. Whereas angled probe A views a particle that has moved less than the alignment particle. This is illustrated in the bottom image of Fig. 2.

Now, imagine multiple time steps are taken. After each time step, the probes each record a new particle compared to the start of the time step. Therefore, the recording from a probe will be of a series of different particles and as each particle has the same velocity–time profile, the traces recorded by the probes will appear elongated and compressed compared to each other. In the example given, the perpendicular and angled B probe will view the end of the cylinder wall faster than the alignment particle travels the cylinder wall, so the recorded trace will appear compressed compared to the particles wall velocity history. Whereas, angled A probe will be slower, so the recorded trace will appear elongated.

4 Reconstruction method

The wall velocity of a particle on the cylinder wall cannot be recorded by a singular HetV probe and instead, so instead it is reconstructed from recordings from multiple probes. This section describes a new approximate reconstruction method that uses the recording from a HetV probe fielded perpendicular to the cylinder wall and another at a known angle. The derivations of the equations used are given in the Appendix.

4.1 Method

The cylinder test HetV traces nominally consist of a curved increase in velocity, with 6–8 peaks and trough pairs in the first ~5 μs that are caused

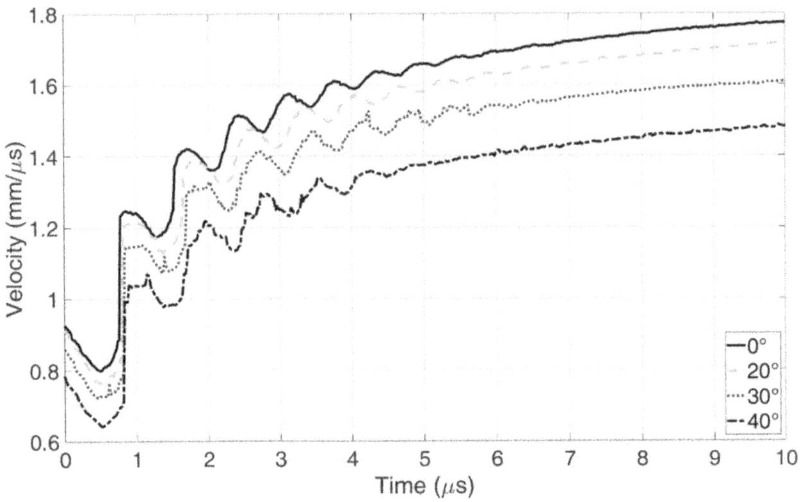

Fig. 3 HetV traces from a cylinder test recorded by probes 0, 20, 30 and 40 degrees to the cylinder wall normal.

by shock wave reflections within the cylinder wall. A typical example set of traces is shown in Fig. 3.

As the traces are recording the steady-state expansion, the turning points are known to occur at the same particle expansion. Therefore, the velocity components recorded by the angled probe (ν_{ap}) and the perpendicular probe (ν_r) at the turning points can be used to calculate the axial component (ν_z) using:

$$\nu_z = \frac{\nu_{ap} - \nu_r \cos(\theta)}{\sin(\theta)} \tag{1}$$

where θ is the angle of the HetV probe relative to the cylinder wall normal. Additionally, ν_z is recorded by the two parallel probes which can be used as an approximate final ν_z value. These values can be linearly interpolated over the perpendicular probe time base (t) to create an approximate axial velocity history.

The time base of the perpendicular probe is different from the wall velocity of a single particle on the cylinder wall as it has been elongated/compressed. Therefore, the perpendicular probe time base (t) is transformed into the time base of a particle on the cylinder wall (T) by:

$$T = \int \frac{D}{D - \nu_z(t)} \mathrm{d}t \tag{2}$$

where D is the detonation velocity. The motion of a particle on the cylinder wall can now be characterised by a plot of (ν_r, T).

4.2 Application

To demonstrate the method, it will be applied to the example data shown in Fig. 4. The turning points were identified in each trace using a peak finding algorithm to identify an initial set of values which were edited by

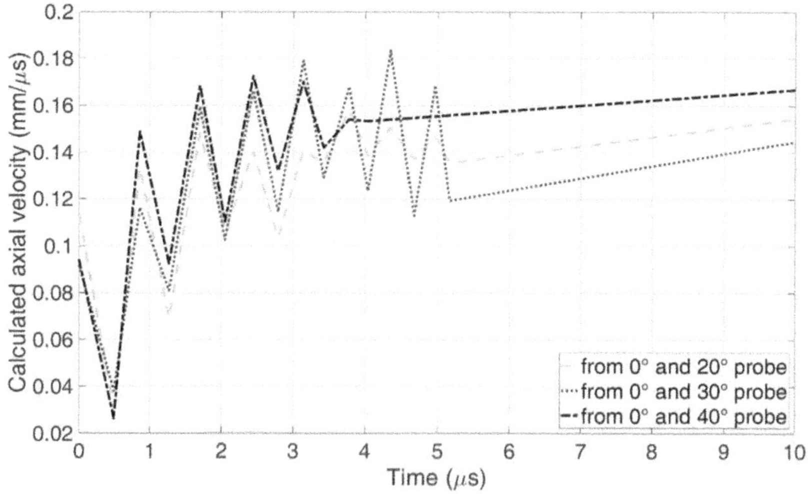

Fig. 4 The three axial velocity traces calculated respectively from the 0° trace and an angled trace.

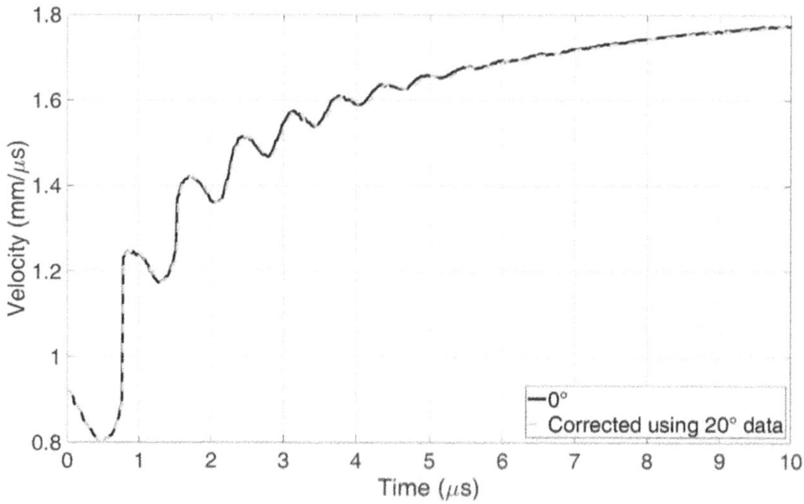

Fig. 5 The velocity trace of the original data recorded by the 0° probe and the data once it has been corrected using the 20° trace.

hand to remove anomalies. These values were used in eqn (1) to create the approximate axial velocity history for the perpendicular probe shown in Fig. 4. For the 0°, 20° and 30° probe, 16 peak/troughs were identifiable (including jump-off) and for the 40° probe, 11 were identifiable, so the axial velocity from the 0° and 40° only uses data from 11 peaks and troughs.

The axial velocity history for each set of probes was numerically integrated in eqn (2) using a cumulative trapezoidal function. The original trace and one example corrected trace are shown in Fig. 5. Fig. 6 shows the percentage increase applied to the time base of the 0° probe data

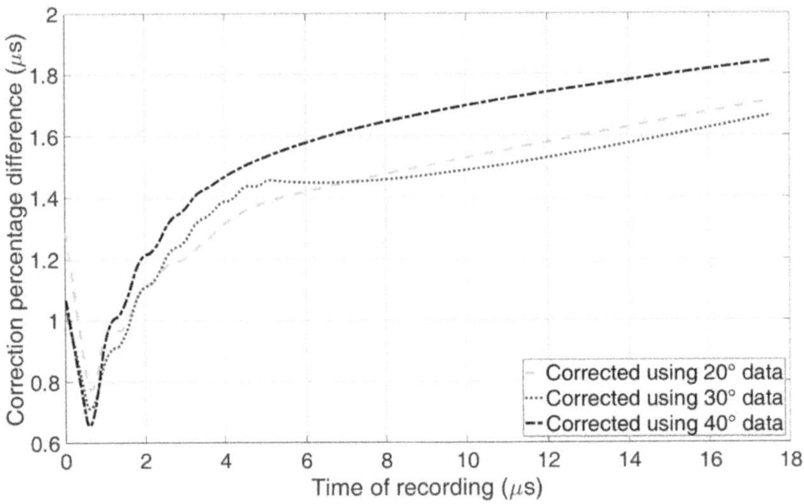

Fig. 6 The percentage difference increase applied to the time of the original data recorded by the 0° probe once it has been corrected against time.

against time for the correction created by the three angled probes. There is good agreement between the three corrected traces and the traces vary between a 0.6 to 1.8% increase.

5 Conclusions

A new approximate technique to correct the data recorded in a cylinder test using two HetV probes fielded perpendicularly and at an angle respectively has been developed and demonstrated. The example data set from an HMX-based explosive has shown that the recorded time base requires a 0.6 to 1.8% increase to match the velocity–time history of a particle on the cylinder wall. This highlights the necessity of a reconstruction method to be applied to HetV cylinder test data to obtain the most accurate Jones–Wilkins–Lee equation of states.

Appendix

A.1 Deriving eqn (1)

Consider a particle on the cylinder wall that travels with the wall velocity at an angle α to the cylinder wall normal, where the wall velocity is expressed by the z and r component ν_z and ν_r respectively. The particle is recorded by the angled and perpendicular probe, where the angled probe is fielded at an angle θ to the cylinder wall normal. The velocity that the angled probe records (ν_{ap}) can be described as a summation of the z and r components of the wall velocity by:

$$\nu_{\text{ap}} = \nu_z \sin(\theta) + \nu_r \cos(\theta) \tag{3}$$

This is illustrated in Fig. 7.

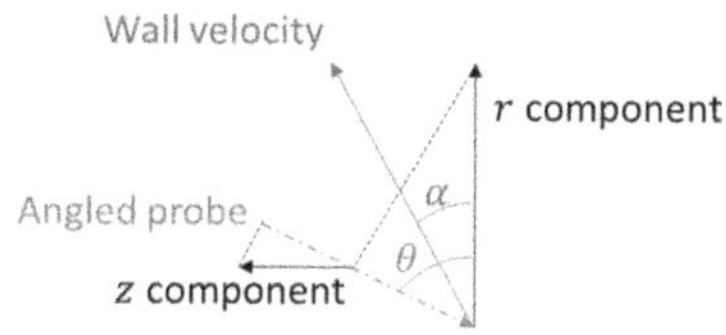

Fig. 7 Illustration of the z and r components of the wall velocity and how they relate to the angled probe recording.

The perpendicular probe records the radial velocity expansion (ν_r), so eqn (3) can be re-arranged to calculate the unknown z component of the wall velocity (ν_z):

$$\nu_z = \frac{\nu_{\text{ap}} - \nu_r \cos(\theta)}{\sin(\theta)} \tag{4}$$

Additionally, the angle of the wall velocity can be calculated from:

$$\tan(\alpha) = \frac{\nu_z}{\nu_r} \tag{5}$$

A.2 Deriving eqn (2)

Consider the reference frame where the detonation wave remains stationary. The HetV probes will move past the cylinder wall with the detonation velocity D, and each particle in the cylinder wall moves through the cylinder wall with the detonation velocity D minus the wall velocity ν. The axial motion of the probe along the cylinder can be described by:

$$z = Dt \tag{6}$$

And the axial motion of the particle can be described by:

$$z = DT - \int \nu_z(T)\mathrm{d}T \tag{7}$$

where t and T is the time the probe and particle have been moving respectively. Therefore, for an arbitrary position on the cylinder wall, the time that the probe records a position (t) will differ from the amount of time the particle has been moving (T). However, a relationship between the two can be derived by equating eqn (8) and (9) that is:

$$t = \int 1 - \frac{\nu_z(T)}{D}\mathrm{d}T \tag{8}$$

or:

$$T = \int \frac{D}{D - \nu_z(t)}\mathrm{d}t \tag{9}$$

References

1 B. D. Sutton, J. W. Ferguson and A. N. Hodgson, *AIP Conf. Proc.*, 2017, **1793**, 030032.

2 S. I. Jackson, *Proc. Combust. Inst.*, 2015, **35**(2), 1997–2004.

3 A. N. Hodgson, J. W. Ferguson and B. D. Sutton, *AIP Conf. Proc.*, 2017, **1793**, 160006.

4 M. Chaos, *Propellants, Explos., Pyrotech.*, 2022, **47**(6), e202100349.

5 O. T. Stand, *26th International Congress on High-Speed Photography and Photonics*, 2005, vol. 5580.

Frictional heating measurements in PBX 9501

Alan Williams,* Judith Brown and Michael Kaneshige

Heating due to frictional processes is one of many proposed mechanisms for hot spot ignition in explosives during sub-shock, mechanical loading. Novel experiments were conducted to isolate this mechanism and measure the temperature rise due to frictional heating in PBX 9501. A rheometer was used to apply mechanical sliding loads at the interface between two explosive pellets with a controlled axial load, and the resulting coefficient of friction of the material sliding against itself and temperature rise using thermocouples along the pellet surface were measured. The measured coefficient of friction values are reported for three different axial loads, and these fall within the range of typical values for polymer against polymer. Measured bulk temperature rises were small, but showed a clear trend with increasing axial load. Such experiments provide valuable data that can be used to inform and/or validate model representations of frictional heating.

1 Introduction

Plastic bonded explosives (PBXs) may be subjected to various mechanical stimuli during their life cycle, including the potential for accidental impact events. Evidence of violent reactions during low-velocity impacts abounds,[1] and this regime of behaviour can be described as mechanically-induced High Explosive Violent Reaction (HEVR). Characterizing the deformation mechanisms and their thermal effects that occur within the explosive during such events is crucial to advance our knowledge of material behavior. Such discovery is needed to inform physics-based models of HEVR, which, in turn, can be used to engineer solutions that mitigate accident severity.

Mechanical deformation of explosives is governed at the mesoscale by several mechanisms, including interfacial debonding between explosive crystals and the polymer binder, crystal fracture, crystal rotations, intracrystalline plasticity, viscoelasticity of the binder, and friction along crack surfaces that may slide against each other during continued deformation. Various subsets or combinations of these mechanisms are hypothesized to contribute to heating of the material, potentially leading to localized hotspot formation and ignition during dynamic, high strain rate loadings.[2,3] Characterizing a single mechanism independently to inform how it should be treated in a HEVR material model, however, is challenging due to the complex material behaviour of these highly filled polymer composite materials.

Frictional heating along internal closed-crack surfaces is an excellent example of one such mechanism that has been hypothesized to contribute to hot spot formation in many models,[3–6] but has been little studied experimentally in PBX. The skid test has historically been used to

Sandia National Laboratories, Albuquerque, NM 87185, USA.
E-mail: alawill@sandia.gov

assess material susceptibility to ignition when friction occurs as the explosive is dragged along a substrate surface with or without grit particles.[7] The heating mechanism in such cases is friction between the grit particle(s) and the hard substrate surface or between the grit particles themselves,[7] whereas heating along internal closed-crack surfaces would be characterized by friction of the PBX material with itself.

The objective of this study is thus to experimentally characterize the thermal effects caused by frictional heating as a PBX material slides against itself in isolation from other deformation mechanisms that could occur simultaneously during a complex impact event. We do this by frictionally sliding two samples of the same PBX 9501 material against each other and measuring the bulk temperature rise. In the following sections, the experimental configuration is described, results are presented to determine the overall temperature rise and coefficient of friction, and a simplified thermal model is used to assess our ability to represent the frictional heat generation process.

2 Experimental

Experiments were conducted to apply mechanical, frictional loads at the interface between two PBX 9501 material surfaces and the thermal response was measured. To do this, two PBX 9501 pellets were mounted concentrically with surfaces touching inside a rheometer, an axial load was applied, then one sample revolved around the other at a set angular sliding speed as shown in Fig. 1.

The resulting normal force and torque were measured and used to calculate the dynamic coefficient of sliding friction (μ) according to eqn (1),[8] where T is the measured torque, F_N is the measured normal force, and R is the effective radius (6.35 mm).

$$\mu = \frac{T}{RF_N} \tag{1}$$

Two PBX 9501 pellets (each 6.35 mm diameter by 6.35 mm height; pressed to 96% of the theoretical maximum density (TMD)) were fixed

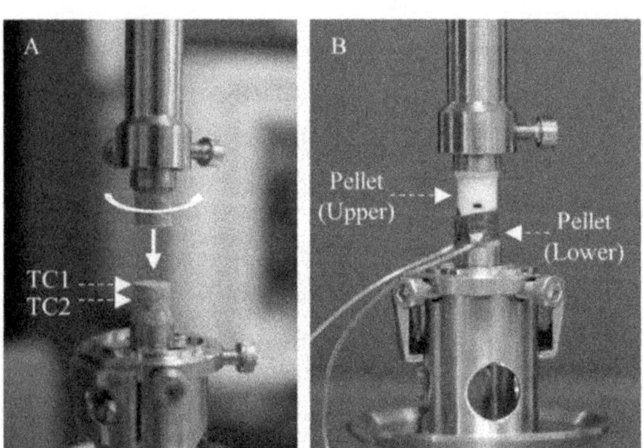

Fig. 1 (A) The mock upper pellet is lowered to contact the lower pellet and revolves to generate heat *via* friction. (B) PBX 9501 samples in contact with two instrumented thermocouples fixed to the lower pellet.

concentrically in a rheometer (TA Discovery Hybrid Rheometer). The rheometer includes an environmental test chamber which provides thermal stability during testing while allowing thermocouple (TC) wiring to pass through. Each pellet is epoxied (Barco Bond MB-185) to a fixture and allowed to set while clamped to preserve concentricity with the fixture. Two T-type thermocouples (Omega® 5SC-TT-T-36-36) were used to measure thermal response and were fixed to the radial surface of the lower pellet using two passes of electrical tape, or one to fix each thermocouple (Fig. 1B). One thermocouple ("TC1") was fixed close to the interface between the pellets at 0.762 mm from the surface and one thermocouple ("TC2") was fixed at 3.048 mm from the surface, or close to the center of the pellet.

Practice tests were performed using density mock samples (900-24; blend of Cyanuric Acid and talc) to determine the ideal loading conditions that achieved a measurable temperature rise while preventing instabilities in the rheometer under higher motion. The procedure for each test was,

1. Fix samples into the Rheometer.
2. Seal the environmental chamber.
3. Wait 10 minutes for thermal equilibrium.
4. Lower the upper pellet onto the lower pellet by setting displacement until the desired set normal force is measured (2.5, 5, or 10 Newtons).
5. Wait 5 minutes after normal load is applied to allow pellets to settle under load, adjusting displacement as necessary until force stabilizes.
6. Run rheometer at 3 radians per second for 180 seconds and collect data.

The axial force was varied at three levels: 5, 2.5, or 10 Newtons. Thermocouple data acquisition was calibrated using a TC simulator calibrator (Ectron Corp 1140A). The rheometer output data (normal force, torque) and TC data was collected at a frequency of 10 Hz. Data was filtered *via* post processing using a moving average filter at a 75-sample span.

3 Results and discussion

3.1 Rheometer test results

The frictional heating tests were run in triplicate repeatability for each of the three loading conditions, where the set axial force was varied at 2.5, 5, or 10 Newtons. The outputs of interest are shown in Fig. 2. Each plot is a representative output for the given loading condition and presents the measured axial force and torque from the rheometer along with the two thermocouple measurements.

The results show an even loading profile across the three conditions, with axial force and torque remaining steady throughout the tests. Both TCs record a measurable temperature rise at the surface of the pellet. As expected, TC1 records a greater temperature rise due to its proximity to

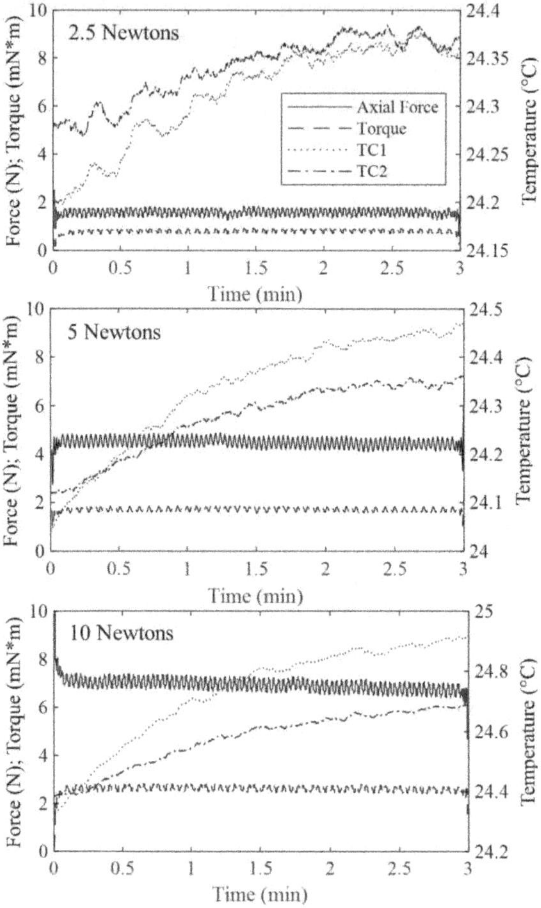

Fig. 2 Representative rheometer outputs of axial force and torque and the temperature rise measured *via* thermocouples for the 2.5 Newton axial loading condition, 5 Newton condition, and 10 Newton condition.

the source of heat generation at the interface between the two pellets. A common feature of this experiment, and tribology testing more generally, is that the applied normal force tends to stabilize over time during the test as the equipment components gradually shift due to test rig compliance. Although this effect is reduced by stabilizing the force in the procedure (steps 4–5), it is inherent and most pronounced for the higher loading conditions (Fig. 2C), where the axial force is initially set at 10 N and falls to approximately 7 N later in the run.

The tabulated results for each loading condition are given in Table 1 and include standard deviations over the three repeat tests and the calculated dynamic coefficient of sliding friction.

Although the coefficient of friction was expected to remain consistent across loading conditions, the 2.5 N loading case showed a significantly higher value with greater variance.

Further investigation here revealed a testing artifact: the experimental stack up (pellets, grips, rheometer arms) are not in perfect alignment

Table 1 Tabulated results for rheometer tests and total temperature rise from thermocouples. Standard deviations are provided over three replicate tests.

Set axial force (N)	Avg. measured force (N)	Avg. coefficient of friction	ΔTC1 (°C)	ΔTC2 (°C)
2.5	2.22 ± 0.891	0.15 ± 0.011	0.23 ± 0.068	0.14 ± 0.026
5	4.57 ± 0.180	0.12 ± 0.001	0.43 ± 0.018	0.27 ± 0.029
10	7.65 ± 1.498	0.12 ± 0.001	0.65 ± 0.114	0.35 ± 0.049

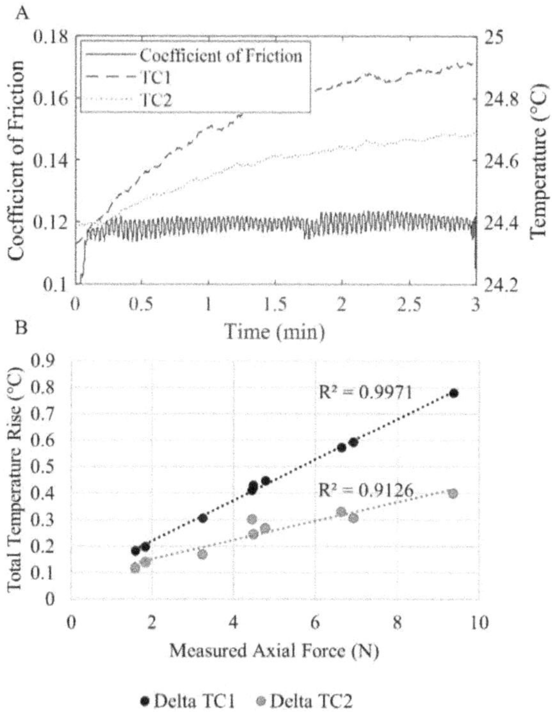

Fig. 3 (A) The calculated coefficient of friction for a representative run (10 Newton loading case) and (B), the total temperature rises for both thermocouples over all measured axial loading conditions.

initially, but rather settle into exact alignment naturally when the axial force is applied. We hypothesize that the lowest axial loading condition (2.5 N), however, was not substantial enough to obtain perfect alignment in our particular equipment, and thus the pellets did not always achieve full surface contact during the 2.5 N tests. Consequently, during part of the sliding oscillation of these tests, the rheometer recorded zero force and torque, leading to a misrepresentation of the coefficient of friction. This zero-force measurement was not observed at the higher axial load conditions (5 N, 10 N), indicating that the pellets maintained full contact throughout the oscillation which leads to a more reliable calculation for the coefficient of friction from these data. This effect can be mitigated in future tests by applying highly precise alignment methods.

An example trace of the coefficient of friction throughout an entire run is given in Fig. 3A. The total temperature rises recorded by both thermocouples are shown in Fig. 3B and are compared to the measured normal force, which more accurately represents the loading state than the set normal force. The total temperature rise exhibits a nearly linear relationship with the increased loading conditions for both thermocouples, particularly for TC1.

3.2 Surface characterization

The frictional heating experiment aims to isolate the mechanism of friction as the primary source of heat generation. To assess the potential contribution of other mechanisms that might occur on the pellet surfaces, such as plastic deformation, crystal pull-out, and fracture, optical microscopy was performed on the contacted surface for one of the pellets. It is expected that visible features such as significant scratching, fracture, pitting from pulled-out crystals, and wear would represent such alternative mechanisms.

As illustrated in Fig. 4, a pristine, untested pellet surface is compared to one of the two sample pellets subjected to loading conditions from all runs. Some minimal surface discoloration is apparent in a spotting pattern which we believe is likely residue from repeated experimental handling. Minor chips are observed along the outer radius and sporadically across the surface.

At higher magnifications (Fig. 4C and D), the surfaces appear nearly identical, with the exception of slight circular scratching patterns on the tested sample. Collectively, these surface features suggest that other deformation-induced heating mechanisms in addition to frictional heating may be present to a small extent, but these are unlikely to significantly contribute to the overall temperature rise.

3.3 Frictional heating model

A thermal model was developed based on the experimental set-up, with the goal to assess how well existing models for frictional heating could represent the heat generation mechanism induced during the test. Only

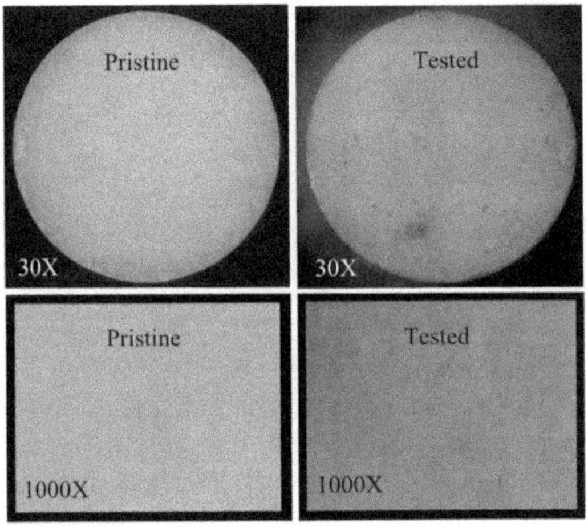

Fig. 4 A pristine PBX 9501 sample compared to a frictionally tested sample subjected to all nine runs of the experiment. Higher magnification (1000×) images show minimal plastic deformation, scratching, and wear on the tested sample's surface.

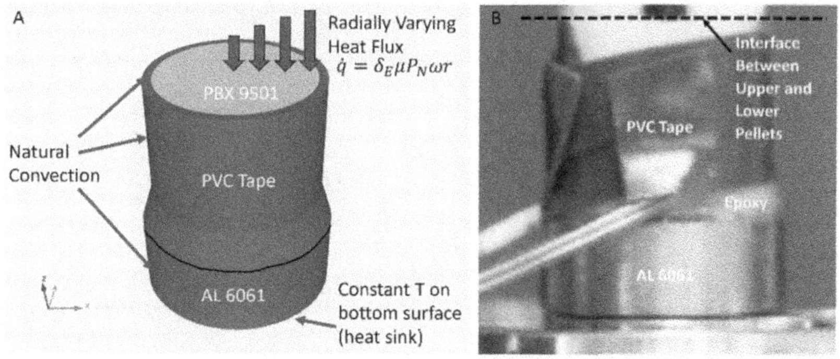

Fig. 5 (A) Thermal model setup (lower PBX 9501 pellet, tape, portion of Al base) and applied boundary conditions, (B) zoomed-in view of lower half of experiment.

the bottom half of the experiment was modelled as shown in Fig. 5, as both thermocouples were attached to the lower pellet and heat flux generated by friction at the pellet interface is assumed to transfer equally into each pellet. Each of the nine tests was modelled separately and results were compared with the respective experiments.

Following a common approach used in friction stir welding models,[9] an applied heat flux boundary condition was used to represent the heat generated by friction at the contacting interfaces of the upper and lower pellets. The prescribed heat flux, $q(r)$, is zero at the centre of the surface ($r = 0$) and varies linearly with radial distance (r) outward as

$$q(r) = \delta_{\mathrm{E}} \mu \frac{F_{\mathrm{N}}}{A} \omega r \tag{2}$$

where μ is the coefficient of friction, F_{N} is the applied normal force, A is the surface cross-sectional area, and ω is the angular velocity of rotation in radians. The pre-multiplier term, δ_{E}, is an efficiency factor used to represent the effectiveness of heat transfer from the surface into the bulk of the pellet.[9] The measured values of friction coefficient, angular rotation frequency, and applied normal force were used to calculate the frictional heat flux for each test. For all of the tests, we assumed an efficiency factor of 0.8.

The outer side surfaces of the model were prescribed with a convective heat transfer coefficient of 50 $\mathrm{W\,m^{-2}\,K^{-1}}$ to approximate the effects of natural convection inside the rheometer's environmental chamber. The bottom surface of the aluminium grip was prescribed a constant temperature corresponding to the initial ambient temperature inside the chamber, to approximate the effect of the large volume of metal apparatus below this acting as a constant temperature heat sink. Thermal material properties used for the model are given in Table 2.

Several additional simplifying assumptions are relevant for the thermal model:

- The electrical tape is assumed to be primarily made from flexible polyvinyl chloride (PVC) polymer. Two layers of tape are assumed to cover the entire pellet side surface with each layer being ~0.01778 mm (0.7×10^{-3} in.) thick—this is modelled as a single

Table 2 Thermal and physical properties used for materials in the model.

Material	Density ($kg\,m^{-3}$)	Thermal conductivity ($W\,m^{-1}\,K^{-1}$)	Specific heat ($J\,kg^{-1}\,K^{-1}$)
PBX 9501	1860 [CF]	0.454 [CF]	1076 [CF]
PVC electrical tape	1225[a]	0.15[a]	900[a]
Al 6061	2712 [CF]	154.5 [CF]	897 [CF]

[a] Estimated based on ranges from ref. 10 and 11.

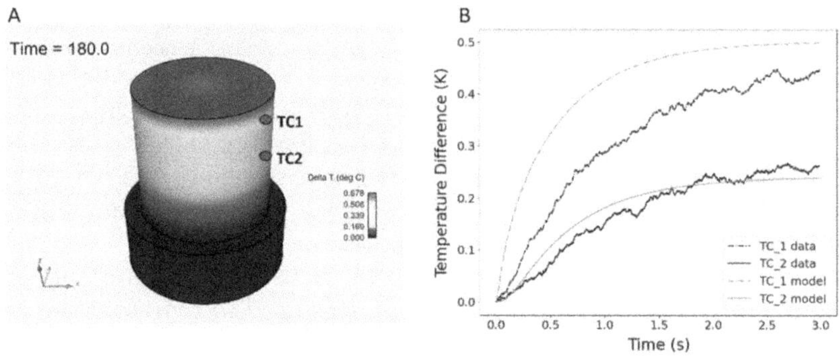

Fig. 6 Model results showing (A) temperature rise contours in the 3D model, (B) temperature *vs.* time curves extracted from the model at locations of TC1 (0.762 mm from pellets interface) and TC2 (3.048 mm from pellets interface) compared with the thermocouple measurements.

layer of tape 0.03556 mm (1.4×10^{-3} in.) thick. As seen in Fig. 5B, the tape does not always cover the whole side surface uniformly; however, we make this assumption for simplicity and consistency in modelling approach for all the tests.

- The frictional heat transfer efficiency factor of 0.8 is estimated to roughly account for the heat loss that would occur at the surface due to uneven tape coverage not represented in the model.
- We neglect the thin layer of epoxy used to attach the PBX 9501 pellet to the aluminium rheometer fixture as the thermal properties of epoxy are similar to those of PBX 9501 and the PVC tape, and this layer is well below the thermocouple measurement locations.
- The thermocouples are not modelled explicitly; rather, the computational results are queried at their specified locations for comparisons.

Fig. 6 shows representative results from the thermal model for one of the tests with 5 N axial loading.

Both the model and the experiment show a concave down trend in temperature rise over time, with the temperature rise levelling off towards a plateau value after ~2 s. The model over-predicts the temperature rise compared to the experimental data at the TC1 location, which is nearest to the frictional heating surface. This is likely due to the insulating effect of the tape, which is assumed to cover this region completely in the model but did not always extend fully to the upper pellet surface in

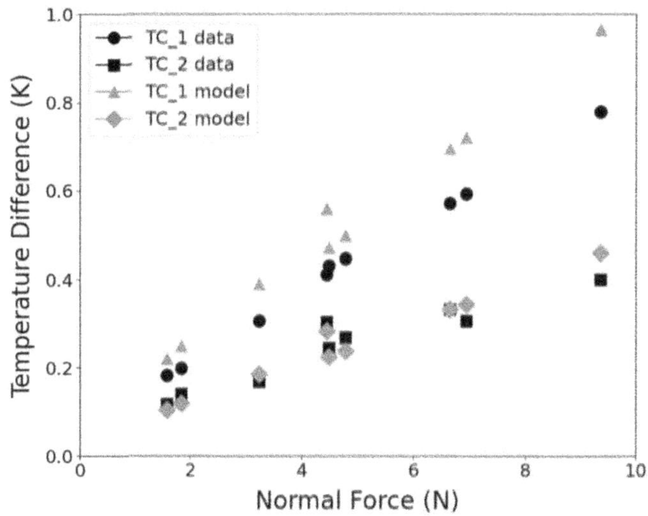

Fig. 7 Measured and predicted temperature rise at the two thermocouple locations as a function of applied normal force.

the tests (see Fig. 5). Other than this effect, the model is in reasonable agreement with the measured thermocouple data.

A summary comparison between the model and experiments of the net temperature rise (time = 3 s) is given for all of the tests in Fig. 7. These values are plotted against the applied normal force from each test in a similar manner to Fig. 3.

From Fig. 7, the model predicts the final temperature rise in good agreement with the measured data across all of the experimental test suite. The model consistently overpredicts the temperature rise at the TC1 location, which we attribute to the assumptions regarding tape coverage of this area. The agreement at the TC2 location, which is further away from the frictional heating surface and more consistently covered by the tape during experiments, is better. These results give confidence that even though there are many simplifications in the model (which are loosely accounted for through the choice of efficiency factor value), the physical mechanism of self-sliding frictional heating in PBX 9501 at these levels of normal force is well represented with a simple linear model of Coulomb friction.

4 Conclusions

In a complex mechanical loading event of a PBX, multiple deformation mechanisms are considered to contribute to the overall heat generation that may lead to the formation of hotspots and thermal runaway. This study investigates a single, isolated mechanism: the frictional heating of PBX 9501 resulting from sliding surface contact.

Experiments were conducted using a TA rheometer to apply frictional loads between two PBX pellets, with one pellet revolving the other at a set angular velocity. As the axial force between the pellets was varied, the increase in temperature at the pellet surface, while modest, was measurable and the coefficient of friction was calculated. A thermal model was developed to complement the experimental observations by predicting the heat generation using a heat flux boundary condition considered in frictional stir welding. The analysis indicates a linear relationship between the applied normal force and the resulting temperature rise, agreeing with the experimental findings. This suggests frictional heating as a significant mechanism contributing to thermal dynamics as a result of mechanical loading.

This study contributes to the understanding of mechanically induced High Explosive Violent Reactions (HEVR). Future work may be extended to include various PBX materials and surface conditions, developing a greater understanding of temperature effects from various material loading conditions and coefficients of friction, which can more completely represent frictional sliding along closed-crack surfaces.

Acknowledgements

The authors are grateful to Hua Wang and Shane Snedigar for testing and experiment execution. Also, thank you to Jason Phillips and Christina Profazi for material characterization. This work was supported by the ASC, DE, and the Laboratory Directed Research and Development program at Sandia National Laboratories. Sandia National Laboratories is a multimission laboratory managed and operated by National Technology & Engineering Solutions of Sandia, LLC, a wholly owned subsidiary of Honeywell International Inc., for the U.S. Department of Energy's National Nuclear Security Administration under contract DE-NA0003525. This paper describes objective technical results and analysis. Any subjective views or opinions that might be expressed in the paper do not necessarily represent the views of the U.S. Department of Energy or the United States Government.

References

1 M. Holmes, G. R. Parker, P. Dickson, E. Heatwole and B. Broilo, *AIP Conf. Proc.*, 2018, **1979**, 150019.
2 J. E. Field, *Acc. Chem. Res.*, 1992, **25**, 489.
3 K. Yang, Y. Wu and F. Huang, *Eur. J. Mech. A Solids*, 2020, **80**, 103924.
4 J. K. Dienes, Q. H. Zuo and J. D. Kershner, *J. Mech. Phys. Solids*, 2006, **54**, 1237.
5 D. Picart, J. Ermisse, M. Biessy, E. Bouton and H. Trumel, *Int. J. Energ. Mater. Chem. Propul.*, 2013, **12**, 487.
6 W. H. Andersen, *Propellants Explos.*, 1981, **6**, 17.
7 G. R. Parker, M. D. Holmes, E. M. Heatwole, R. M. Broilo, M. N. Pederson and P. M. Dickson, *Combust. Flame*, 2020, **221**, 180.
8 J. L. Meriam and L. G. Kraige, *Engineering Mechanics: Dynamics*, John Wiley & Sons, 7th edn, 2012.
9 C. Hamilton, S. Dymek and A. Sommers, *Int. J. Mach. Tools Manuf.*, 2008, **48**, 1120.
10 B. H. O. Mulina, V. L. Borges, A. G. Trovo and S. R. Carvalho, *J. Braz. Soc. Mech. Sci. Eng.*, 2022, **44**, 44.
11 R. Bhowmik, S. Sihn, V. Varshney, A. K. Roy and J. P. Vernon, *Polymer*, 2019, **167**, 176.

Subject Index